Blackstone's Police Manual

Road Traffic

Blackstone's

Police Manual

Road Traffic

2004 edition

Fraser Sampson

LLB, LLM, MBA, Solicitor

OXFORD
UNIVERSITY PRESS

Great Clarendon Street, Oxford OX2 6DP

Oxford University Press is a department of the University of Oxford.
It furthers the University's objective of excellence in research, scholarship,
and education by publishing worldwide in

Oxford New York

Auckland Bangkok Buenos Aires Cape Town Chennai
Dar es Salaam Delhi Hong Kong Istanbul Karachi Kolkata
Kuala Lumpur Madrid Melbourne Mexico City Mumbai Nairobi
São Paulo Shanghai Taipei Tokyo Toronto

and an associated company in Berlin

Oxford is a registered trade mark of Oxford University Press
in the UK and in certain other countries

Published in the United States
by Oxford University Press Inc., New York

A Blackstone Press Book

First published 1998
Reprinted with amendments, November 1998
Second edition, 1999
Third edition, 2000
Fourth edition, 2001
Fifth edition, 2002
Sixth edition, 2004

British Library Cataloguing in Publication Data
Data available

Library of Congress Cataloging in Publication Data
Data available

ISBN 0-19-9262454

1 3 5 7 9 10 8 6 4 2

Typeset by Newgen Imaging Systems (P) Ltd., Chennai, India
Printed in Great Britain
on acid-free paper by
Ashford Colour Press, Gosport, Hampshire

Contents

Foreword for 2004 Blackstone's Police Manuals

In the 21st century, police officers must demonstrate a wide range of knowledge, skills, and abilities – not only if they are to perform effectively in their current role, but also if they are seeking to gain promotion to higher ranks within the service. The knowledge and understanding of relevant law and procedure remains central to the role of any officer, as does the ability to effectively apply this law and procedure in their day-to-day duties.

The Blackstone's Police Manuals are presented in four volumes – crime, evidence and procedure, road traffic and general police duties. The manuals have been fully updated to reflect recent changes in legislation, including the 2003 revisions to the PACE Codes of Practice. Oxford University Press work alongside Centrex to ensure that the books are fully accurate and up-to-date, and reflect the content which is required by a police officer in today's service.

The Blackstone's Police Manuals are primarily designed to assist candidates preparing for the Sergeants' and Inspectors' OSPRE Part I Police Promotion Examination. In serving this purpose, the books should be used in combination with the official OSPRE Rules & Syllabus document. You may have received a copy of this document when you obtained these manuals. However, additional copies of the Rules & Syllabus can be downloaded from the Employers' Organisation for Local Government web site at www.lg-employers.gov.uk.

From a wider perspective, the Manuals are also designed to provide an ongoing reference point for officers seeking to keep their professional knowledge up-to-date. Moreover, the Manuals continue to be widely used by officers seeking a concise explanation of specific aspects of the law they use in their day-to-day duties.

Centrex continues to play a key role within the modern police service through assisting officers in their ongoing professional development. The OSPRE system, developed and delivered by Centrex Examinations & Assessment, not only selects tomorrow's managers and supervisors, but also assists candidates in developing themselves professionally in the knowledge and skills required for these ranks. I believe the Blackstone's Police Manuals continue to play a central role in this process.

If you are using the books to prepare for the OSPRE Promotion Examinations, may I take this opportunity to wish you the best of luck in your studies, and I hope the books will assist you in progressing your career within the service.

Chris Mould
Centrex
Chief Executive
August 2003

Preface

The pressure to develop and apply occupational standards continues for the police. In moving towards these standards, the service is having to revisit many of the old debates about the depth and breadth of legal knowledge that can be properly expected of its officers and staff. Police staff of all ranks and position are expected—not unreasonably—by those they serve and by the courts to know the law. However, as the additions to all of the Manuals in this series show, the area that is now becoming recognised as 'police law' is being made and implemented at a dizzying pace (no pun intended). By and large, the responsibility for keeping abreast of these legal developments is still left to the individual—and, in trying to keep up, individuals with the police service turn to *Blackstone's Police Manuals* as their first port of call.

The leading reference source on police law, and used by officers of all ranks, as well as lawyers, advisers—and even the odd judge—the Manuals have grown considerably in popularity as well as size. Plans to make the Manuals available in electronic format and to supplement them with other operationally-focused materials mean that they will become more accessible and comprehensive than ever, providing an invaluable policing tool for the future.

The Manuals are fully indexed and cross-reference and are published each year to keep them up to date, doing away with the need for inserts or supplements. More importantly for OSPRE candidates, if the law isn't in the latest edition of the Manual, it won't be in the exam.

Covering the key principles and the content of the various syllabuses, this Manual is intended to navigate readers through the oceans of legal regulation, rather than drown them in its volume. The idea is to include enough relevant detail to provide a proper legal reference source, while avoiding the minutiae that characterises this most pedantic of subjects.

And if police law in general is pedantic, then road traffic as a sub-set is positively finical. Tackling any subject which covers police drivers jumping red lights, reverse burdens of proof, minimum ages for driving lawnmowers, the precise dimensions of characters on number plates and the distances from which they must be read and Pelicans, Puffins and Toucans calls for some judicious pruning. As with 'Best Man' speeches, in writing books like this you have to give as much thought to what to leave out as what to put in and you are still bound to upset someone.

While every care has been taken to ensure the accuracy of the contents of this Manual, neither the author nor the publishers can accept any responsibility for any actions taken, or not taken, on the basis of the information contained in this Manual.

The law is stated as at June 2003.

Acknowledgements

Blackstone's Police Manuals have become firmly established as a 'household name' in the context of police law for which they are the leading text in England and Wales. Their growth and refinement year on year is a result, not simply of Parliament and the legislators, but also of the many and varied contributions of the Manuals' wide readership.

Thanks are due to all (except perhaps the legislators) and there is never space to mention everyone individually.

Thanks go to all the production and editorial team—Jane (x2), Roxanne, Marianne and Mandy. And, as ever, to CC, TA, AC and now MO for being around when needed even if I wasn't. Special thanks to Sgt Joyce of the Hertfordshire Constabulary.

Fraser Sampson
Walker Morris

Table of Cases

Table of Statutes

International legislation

EC legislation

EC secondary legislation

Table of Secondary Legislation

1 Classifications and concepts

1.1 Introduction

The majority of the most common road traffic definitions can be found in ss. 185 and 192 of the Road Traffic Act 1988. However, there are some other definitions which are specifically set out in the relevant piece of legislation, the most obvious examples being those relating to the construction and use of road vehicles (**see chapter 9**) and driving licences (**see chapter 11**).

There are also some important practical generalisations set out in Part I of the Road Traffic Offenders Act 1988 relating to proof of certain issues—such as the presumed driver of a vehicle at certain times, as well as matters relating to construction and use of vehicles and disqualification of drivers. These latter issues are more of a consideration for the relevant court hearing the matter in question. The definitions that follow in this chapter, however, are critical to police officers when dealing with road traffic offences.

The definition of a particular vehicle will generally be determined by its size, weight and construction; on occasions (as with recovery vehicles—**see para. 1.2.6**), it will also be determined by the way in which the vehicle is used.

In approaching any situation involving road traffic legislation it is critical to establish the exact definition *as it applies to that particular piece of legislation.*

This chapter sets out the more commonly-encountered definitions which are referred to, where appropriate, throughout the rest of this book. Where different definitions apply, they are included in the relevant chapter. For instance, 'large goods vehicles' and 'passenger carrying vehicles' as defined under s. 121 of the Road Traffic Act 1988 are dealt with in **chapter 13**.

1.2 The Definitions

Most of the more commonly-used definitions can be found in the Road Traffic Act 1988 but many are duplicated in other legislation (e.g. the Road Vehicles (Construction and Use) Regulations 1986, SI 1986/1078, as amended).

1.2.1 Motor Vehicles and Mechanically Propelled Vehicles

The definition of a 'motor vehicle' which will apply in most cases can be found in s. 185 of the Road Traffic Act 1988. Section 185(1) defines a motor vehicle as a mechanically propelled vehicle intended or adapted for use on roads. (Similar definitions of motor vehicles, motor cars, etc. can be found under the Road Traffic Regulation Act 1984, s. 136 and the Road Vehicles (Construction and Use) Regulations 1986, SI 1986/1078, reg. 3.)

If a vehicle is not intended or adapted for use on a road it will not be a motor vehicle for these purposes. Whether a vehicle is so intended or adapted is a question of fact to be decided in each case.

But, just in case this concept became a little too straightforward, Parliament has recently introduced two further definitions of motor vehicle for specific purposes. The first applies to s. 59 of the Police Reform Act 2002, where motor vehicle means any mechanically propelled vehicle *whether or not it is intended or adapted for use on roads* (s. 59(9)). This is probably because the power itself can be used where there are reasonable grounds for believing that a vehicle has been used in a way that contravenes several specific offences. Those offences themselves use the definition of a 'mechanically propelled vehicle' but why then the Police Reform Act 2002 could not just use that expression as well is far from clear. For a full discussion of the powers under s. 59, **see chapter 2**. The second new definition of motor vehicle can be found in the Vehicles (Crime) Act 2001. Motor vehicle here *means any vehicle whose function is or was to be used on roads as a mechanically propelled vehicle.*

'Vehicle' for the purposes of the Vehicle Excise and Registration Act 1994 (**see chapter 12**) means a mechanically propelled vehicle or *anything* (whether or not it is a vehicle) that has been, but has ceased to be, a mechanically propelled vehicle (s. 62).

In practice, the various finer points of distinction in these overlapping definitions will not cause too many problems. It is essential, however, from an evidential point of view, to ensure that the vehicle concerned falls within the statutory definition that applies to the particular offence you are dealing with. It is then up to the courts to decide whether that definition does in fact cover the vehicle itself—and there have been a few oddities recently (see below). And finally, having added to the confusion, the Vehicles (Crime) Act 2001 also introduces a new definition, that of 'written-off motor vehicles'. These are motor vehicles in need of substantial repair but in relation to which a decision has been made not to carry out the repairs. For a full discussion of the Vehicles (Crime) Act 2001, **see chapter 12**.

A mechanically propelled vehicle is, quite simply, a vehicle that is constructed so that it can be propelled mechanically. If the vehicle is so constructed, it does not cease to be 'mechanically propelled' just because the mechanism is not being used to propel it at the relevant time. Therefore a motor-assisted pedal cycle (but not an *electrically* assisted one; see below) has been held to be a mechanically propelled vehicle even when it is being pedalled (*Floyd* v *Bush* [1953] 1 WLR 242). So the test as to whether a vehicle is 'mechanically propelled' is one of construction rather than use (see *McEachran* v *Hurst* [1978] RTR 462). The prosecution bears the burden of showing that a vehicle meets the requirements of being 'mechanically propelled' (*Reader* v *Bunyard* [1987] RTR 406).

The test will be an objective one which looks at the *raison d'être* of the vehicle's design; it does not look at the intention of the owner (*Chief Constable of Avon and Somerset* v *Fleming* [1987] RTR 378). Developments in motorised and mechanical propulsion have thrown up a few technical questions for the courts recently. For instance, a petrol-driven micro-scooter was held to be a motor vehicle within the definition of the Road Traffic Act 1988 by magistrates. In that case (*Chief Constable of North Yorkshire Police* v *Saddington* [2001] RTR 227) the Divisional Court upheld the magistrates' decision asking whether a reasonable person would say that one of the 'Go-Ped's' uses would be some general use on the road. Although acknowledging that users would therefore need a driving licence and insurance (as to which, **see chapters 6 and 11**), the court declined to classify 'Go-Peds' in any particular category. However, in *Winter* v *DPP* [2003] RTR 14, the Administrative Court acknowledged a 'City Bug' conveyance as being a mechanically propelled vehicle. The Court went on to uphold the Crown Court's decision that such a contraption could not be made into an 'electrically assisted cycle' simply by the addition of pedals by which the conveyance could not really be propelled and therefore the user required insurance for it when used on a road or other public place.

Note that a motor vehicle continues to be such if it is towed by another vehicle (*Cobb* v *Whorton* [1971] RTR 392), and a car which is rebuilt for off road racing continues to be 'intended' for use on a road even though those rebuilding it never intended to use it so again (*Nichol* v *Leach* [1972] RTR 476).

Dumper trucks intended for use solely on construction sites will probably not be 'motor vehicles' but, if you are able to adduce evidence that they *are suitable* for use on a road, they may be held to be motor vehicles (*Daley* v *Hargreaves* [1961] 1 All ER 552).

'Mechanically propelled' generally includes steam and electrically powered vehicles (*Elieson* v *Parker* (1917) 81 JP 265). It does not include electrically assisted *pedal cycles* (**see chapter 15**). However, in a Scottish case, 'mechanically propelled' has been held to include a van powered by an electric motor with onboard storage batteries (*McMorrow* v *Vannet* 1998 SLT 1171).

Removing an engine from a vehicle does not stop it from being 'mechanically propelled' if you can show that the engine could easily be replaced (*Newberry* v *Simmonds* [1961] 2 All ER 318) or, in a case like *McMorrow* above, by taking the batteries out. A car with a flat battery is still a mechanically propelled vehicle (*R* v *Paul* [1952] NI 61).

Road Traffic Act 1988, s. 189 specifies which vehicles will *not* be treated as motor vehicles; they include pedestrian controlled vehicles and some implements for cutting grass.

1.2.2 Passenger Vehicles

A 'passenger vehicle' is defined under reg. 3(2) of the Road Vehicles (Construction and Use) Regulations 1986 (SI 1986/1078) as being a vehicle 'constructed solely for the carriage of passengers and their effects'.

The test to be applied to a vehicle in order to see if it falls within the definition of a 'passenger vehicle' is by asking whether or not at the material time the vehicle was of a type that would ordinarily be used for the carriage of passengers and their effects (*Flower Freight Co. Ltd* v *Hammond* [1963] 1 QB 275). The inclusion of the word 'solely' in the definition means that vehicles ordinarily used for the carriage of passengers and their effects *or* the carriage of goods will not be 'passenger vehicles' (*Flower Freight Co. Ltd*).

If a passenger vehicle is constructed or adapted to carry more than 8 but no more than 16 passengers it will be a 'mini-bus', and if more than 16 seated passengers it will be a 'large bus' (reg. 3(2)).

References to passengers do not include the driver for this definition.

A large bus as defined above which has a maximum gross weight of 7.5 tonnes and having a maximum speed exceeding 60 mph is a 'coach'. (For large passenger carrying vehicles, **see chapter 13**.)

1.2.3 Goods Vehicles

Although some goods vehicles will also fall under the categories above (e.g. motor cars), 'goods vehicles' are themselves defined under s. 192(1) of the Road Traffic Act 1988 as motor vehicles constructed or adapted for use for the carriage of goods or a trailer so constructed or adapted. 'Goods' for these purposes includes goods or burden of any description. Goods vehicles are further categorised into 'large' goods vehicles and 'medium-sized' goods vehicles. For large goods vehicles, **see chapter 13**. Medium-sized goods vehicles are defined under s. 108(1) of the 1988 Act as motor vehicles constructed or adapted to carry or haul goods, which are not adapted to carry more than nine persons including the driver and the permissible maximum weight of which exceeds 3.5 tonnes but not 7.5 tonnes (including a combination of such a motor vehicle and a trailer where the relevant maximum weight of the trailer does not exceed 750 kg).

1.2.4 Motor Cycles, Motor Bicycles and Mopeds

Confusingly there are differences in definition between motor cycles, motor bicycles, large motor bicycles and mopeds. In practice, the same machine may fall into a number of these definitions. Again, it is important to identify in what context the definition is used.

Motor Cycle

The general definition of a *motor cycle* can be found—alongside most of the other important general definitions—under s. 185(1) of the Road Traffic Act 1988. Section 185(1) defines a motor cycle as 'a mechanically propelled vehicle, not being an invalid carriage, with less than four wheels and the weight of which unladen does not exceed 410 kilograms'. This definition would include some three-wheeler cars and pedestrian controlled vehicles.

For the purposes of part III of the 1988 Act—the part that deals with licensing (**see chapter 11**)—s. 108 creates further definitions of:

- A motor bicycle.
- A moped.

Motor Bicycle

Under s. 108(1) a motor bicycle is defined as:

> ...a motor vehicle which—
>
> (a) has two wheels, and
>
> (b) has a maximum design speed exceeding 45 kilometres per hour and, if powered by an internal combustion engine, has a cylinder capacity exceeding 50 cubic centimetres, and includes a combination of such a motor vehicle and a side-car,...

Moped

Also under s. 108(1), a moped is defined as:

> ...a motor vehicle which has fewer than four wheels and—
>
> (a) in the case of a vehicle the first use (as defined in regulations made for the purpose of section 97(3)(d) of this Act) of which occurred before 1st August 1977, has a cylinder capacity not exceeding 50 cubic centimetres and is equipped with pedals by means of which the vehicle is capable of being propelled, and
>
> (b) in any other case, has a maximum design speed not exceeding 50 kilometres per hour and, if propelled by an internal combustion engine, has a cylinder capacity not exceeding 50 cubic centimetres,...

These definitions are really sub-divisions of motor cycles and are relevant for the purposes of driver licensing.

Learner Motor Bicycle

Section 97 of the Road Traffic Act 1988 deals with the granting of driving licences (as to which, **see chapter 11**). In particular, s. 97 sets out the requirements for, and restrictions on, provisional licences and, in that context, makes reference to *learner* motor bicycles, introducing yet another sub-division of motor cycles. A learner motor bicycle is defined at s. 97(5) as a motor bicycle which is either:

- propelled by electric power; or
- the cylinder capacity of its engine does not exceed 125 cc and the maximum net power output of its engine does not exceed 11 Kw.

For the purposes of driving licence categories, learner motor bicycles fall under category A1 (**see appendix 6**).

Large Motor Bicycle and Standard Motor Bicycle

As if all this were not confusing enough, the Motor Vehicles (Driving Licences) Regulations 1999 (SI 1999/2864) (as to which, **see chapter 11**) create another sub-division; that of 'large' and 'standard' motor cycles.

Large motor bicycles are defined (under reg. 3) as being:

- in the case of a motor bicycle without a sidecar, a bicycle the engine of which has a maximum net power output exceeding 25 Kw or which has a power to weight ratio exceeding 0.16 Kw per kilogram; or
- in the case of a motor bicycle and sidecar combination, a combination having a power to weight ratio exceeding 0.16 Kw per kilogram.

For the purposes of reg. 3, a 'standard' motor bicycle is a motor bicycle that does not fit into the above category (i.e. it is the opposite of a 'large' motor bicycle).

Minimum Ages

The table of minimum ages for holding or obtaining licences in respect of certain vehicles is set out at s. 101 of the Road Traffic Act 1988 (**see appendix 1**). Regulation 9 of the 1999 Regulations increases the minimum age for motor bicycles if the vehicle is a 'large motor bicycle' (except, as you might have guessed, in certain circumstances). That age is increased to 21 years for most cases. The main exceptions are:

- where a person has passed a test on/after 1 January 1997 for a motor bicycle other than a learner motor bicycle (category A) and the 'standard access period' (generally two years) has elapsed; and
- where the large motor bicycle is owned by the Secretary of State for Defence or is being driven by someone who is at that time subject to the orders of a member of the armed forces.

The definition of a 'large' motor bicycle is also relevant for the purposes of a full licence being used as a provisional entitlement to drive some other vehicles (**see chapter 11**).

Motor Tricycles and Quadricycles

Finally, there is a classification of motor tricycles and quadricycles. These are mainly relevant for the purposes of registration marks (see the Road Vehicles (Display of Registration Marks) Regulations 2001 (SI 2001 No. 561), **see chapter 12**). Motor tricycles are vehicles having three wheels symmetrically arranged. Quadricycles are four-wheeled vehicles having a net engine power not exceeding 15 Kw and an unladen mass not exceeding 400 kilograms (unless a goods vehicle in which case the relevant mass is 550 kg). Basically they are quad bikes.

Summary

So, to summarise, the definition of a motor cycle, which will be of general application, can be found under the Road Traffic Act 1988, s. 185.

In relation to the *licensing of drivers of vehicles under part III of that Act*, there are further classifications of motor bicycles and mopeds. These are defined at s. 108 of the 1988 Act.

With regard to the *licensing of learner drivers generally*, s. 97 of the 1988 Act refers to a further classification, namely a learner motor bicycle.

In relation to the *minimum ages for holding and obtaining a driving licence*, and *the provisional entitlements of other licences*, the 1999 Regulations make reference to 'large' motor bicycles and their opposites, 'standard' motor bicycles.

And finally, for the purposes of registration plates, there are further categories of motor tricycles and quadricycles which are relevant when considering number plates.

Having considered all these, it is difficult to look at a motor bike in the same way again!

1.2.5 Trailer

A 'trailer' is defined, for most purposes, under s. 185(1) of the Road Traffic Act 1988 as a vehicle drawn by a motor vehicle (however, a different definition applies under Regulation EEC 3820/85 in relation to drivers' hours).

When a motor vehicle is towed by another motor vehicle, the latter still falls within the category of motor vehicle but can also be regarded as a 'trailer' (*Milstead* v *Sexton* [1964] Crim LR 474).

As 'vehicle' is not defined in the Road Traffic Act 1988 virtually anything would appear to be capable of amounting to a vehicle under the right circumstances, even a chicken shed on wheels (see *Garner* v *Burr* [1951] 1 KB 31) or a hut (*Horn* v *Dobson* 1933 JC 1).

Where a 'heavy motor car' or 'motor car' (not being an articulated bus) draws a trailer with at least 20 per cent of the trailer's load being borne by the drawing vehicle, the combination will be regarded as an 'articulated vehicle' (Road Vehicles (Construction and Use) Regulations 1986, reg. 3(2)).

1.2.6 Other Definitions

Motor Car

Section 185(1) of the Road Traffic Act 1988 also defines a 'motor car':

> . . . 'motor car' means a mechanically propelled vehicle, not being a motor cycle or an invalid carriage, which is constructed itself to carry a load or passengers and the weight of which unladen—
>
> (a) if it is constructed solely for the carriage of passengers and their effects, is adapted to carry not more than seven passengers exclusive of the driver and is fitted with tyres of such type as may be specified in regulations made by the Secretary of State, does not exceed 3050 kilograms,
>
> (b) if it is constructed or adapted for use for the conveyance of goods or burden of any description, does not exceed 3050 kilograms, or 3500 kilograms if the vehicle carries a container or containers for holding for the purposes of its propulsion any fuel which is wholly gaseous at 17.5 degrees Celsius under a pressure of 1.013 bar or plant and materials for producing such fuel,
>
> (c) does not exceed 2540 kilograms in a case not falling within sub-paragraph (a) or (b) above,

However, in relation to the law regulating driving instruction (part V of the Road Traffic Act 1988) there is a further definition of a 'motor car'. This definition is a motor vehicle other than an invalid carriage or motor cycle which is not constructed or adapted to carry more than nine persons including the driver and which has a maximum gross weight not exceeding 3.5 tonnes (s. 141A).

Heavy Motor Car

Section 185(1) defines a heavy motor car as:

> . . . a mechanically propelled vehicle, not being a motor car, which is constructed itself to carry a load or passengers and the weight of which unladen exceeds 2540 kilograms,

Small Vehicle

For the purposes of driver licensing (part III of the Road Traffic Act 1988), a small vehicle is defined as a motor vehicle other than an invalid carriage, motor bicycle or moped, which is not constructed/adapted to carry more than nine persons including the driver and which has a maximum gross weight not exceeding 3.5 tonnes (and including a combination of such a vehicle and a trailer) (s. 108(1)).

Locomotives

Section 185(1) also defines locomotives as:

> 'heavy locomotive' means a mechanically propelled vehicle which is not constructed itself to carry a load other than any of the excepted articles and the weight of which unladen exceeds 11690 kilograms,

> 'light locomotive' means a mechanically propelled vehicle which is not constructed itself to carry a load other than any of the excepted articles and the weight of which unladen does not exceed 11690 kilograms but does exceed 7370 kilograms,

The 'excepted articles' referred to in s. 185(1) are water, fuel, accumulators and other equipment used for the purpose of propulsion, loose tools and loose equipment (s. 185(2)).

Motor Tractor

The definition of a motor tractor can be found, unsurprisingly, in s. 185(1), which states:

> ... 'motor tractor' means a mechanically propelled vehicle which is not constructed itself to carry a load, other than the excepted articles, and the weight of which unladen does not exceed 7370 kilograms,

The 'excepted articles' are those listed under locomotives above.

Invalid Carriage

An invalid carriage is defined (under s. 185(1)) as being:

> ... a mechanically propelled vehicle the weight of which unladen does not exceed 254 kilograms and which is specially designed and constructed, and not merely adapted, for the use of a person suffering from some physical defect or disability and is used solely by such a person,

An invalid carriage complying with the regulations under the Chronically Sick and Disabled Persons Act 1970 will not be treated as a motor vehicle under the Road Traffic Act 1988.

Heavy Commercial Vehicle

A 'heavy commercial vehicle' is defined under s. 138 of the Road Traffic Regulation Act 1984 (and also under s. 20 of the Road Traffic Act 1988) as any goods vehicle with an operating weight exceeding 7.5 tonnes.

Recovery Vehicle

A 'recovery vehicle' is defined under the Vehicle Excise and Registration Act 1994, sch. 1, para. 5:

> ... 'recovery vehicle' means a vehicle which is constructed or permanently adapted primarily for any one or more of the purposes of lifting, towing and transporting a disabled vehicle.
>
> (3) A vehicle is not a recovery vehicle if at any time it is used for a purpose other than—
> (a) the recovery of a disabled vehicle,
> (b) the removal of a disabled vehicle from the place where it became disabled to premises at which it is to be repaired or scrapped,
> (c) the removal of a disabled vehicle from premises to which it was taken for repair to other premises at which it is to be repaired or scrapped,
> (d) carrying fuel and other liquids required for its propulsion and tools and other articles required for the operation of, or in connection with, apparatus designed to lift, tow or transport a disabled vehicle, and
> (e) any purpose prescribed for the purposes of this sub-paragraph by regulations made by the Secretary of State.

The equipment or apparatus for lifting, towing or transporting disabled vehicles must be permanently mounted or form an integral part of the vehicle. Therefore, a vehicle, which was in effect a mobile workshop that had towing and lifting capabilities, was not a recovery vehicle as defined above (*Vehicle Inspectorate* v *Richard Read Transport Ltd*, [1998] RTR 288).

1.2.7 Key Concepts

Weights

The weight of a vehicle will often determine its classification and the use to which it may be put, together with the licensing conditions of those who drive it.

Reference will usually be made to the *laden* weight or the *gross* weight of a vehicle in offences involving its construction and use, while offences concerning the speed and excise duty of larger vehicles will usually be concerned with *laden* and *unladen* weight.

Calculation of a vehicle's *unladen* weight for most purposes is contained under s. 190 of the Road Traffic Act 1988.

Most locomotives, motor tractors and heavy motor cars are required to show their *unladen* weight, either on their nearside or on the relevant plating certificate(s) (see Road Vehicles (Construction and Use) Regulations 1986, regs 66 and 71).

For construction and use purposes, the *gross* weight under reg. 3(2) of the 1986 Regulations will be:

- For motor vehicles—the sum of the weights transmitted to the road surface by all its wheels.
- For trailers—the sum of the weights transmitted to the road surface by all its wheels *and* of any weight of the trailer which is imposed on the drawing vehicle.

Axle weights are, generally, the sum of the weights transmitted to the road surface by all the wheels of that axle.

Accident

'Accidents' are relevant in a number of road traffic offences (e.g. drink driving (**see chapter 5**) and notices of intended prosecution (**see chapter 3**)).

Apart from 'reportable' accidents (**see chapter 4**), there is no single definition of what an accident is. Rather, the courts have warned against creating any hard and fast rule as to what will amount to an 'accident' (*Chief Constable of West Midlands Police* v *Billingham* [1979] RTR 446).

The general test which courts will apply is whether an ordinary person would say that there had been an accident in all the circumstances.

Such an approach will include, not only an unintended occurrence or an inadvertent act, but also a deliberate act by a driver such as ramming a gate (*Chief Constable of Staffordshire* v *Lees* [1981] RTR 506).

There is no need for the involvement of another vehicle in an 'accident' (*R* v *Pico* [1971] RTR 500).

Driver

It is possible for more than one person to be the 'driver' of a vehicle: s. 192(1) of the Road Traffic Act 1988 makes provision for the 'steersman' of a motor vehicle also to be included in the definition of 'driver' (except for the offence of causing death by dangerous driving under s. 1). Where one person sits in the driving seat while another reaches across them and operates the steering wheel, both can be 'drivers' of the vehicle (*Tyler* v *Whatmore* [1976] RTR 83). Similarly, a learner driver and his or her instructor may be the drivers of a vehicle (*Langman* v *Valentine* [1952] 2 All ER 803).

The only situation where the above will not apply is in cases of causing death by dangerous driving under s. 1 of the Road Traffic Act 1988 (**see chapter 2**).

Whether a person was actually *the* driver of a vehicle at a particular time cannot be inferred from the fact that they are also the owner; more evidence of their actual involvement in the driving of the vehicle will be required (*R* v *Collins* [1994] RTR 216).

For the purposes of many road traffic offences, the person who takes out the vehicle remains the 'driver' of it until he/she has finished that journey. Therefore, even if the vehicle is stationary and has been for some time, a person may still be the 'driver' of it if he/she has not completed the journey (see *Jones* v *Prothero* [1952] 1 All ER 434 where the driver sitting in a parked car opened the offside door causing injury to a passing pedal cyclist).

This definition is particularly relevant in cases involving the duty placed on 'drivers' after a reportable accident (as to which, **see chapter 4**). In such cases there appears to be a distinction between a person being a *driver* and someone who is actually engaged in *driving*. Under the legislation relating to accidents, the 'driver' of a vehicle is under an obligation in relation to that vehicle while it remains on a road and that duty can continue, even after the person leaves the vehicle for a short time (see *Cawthorn* v *DPP* [2000] RTR 45; driver's vehicle rolled downhill while he went to use a post box).

Driving

Although s. 192 of the Road Traffic Act 1988 provides that a 'steersman' can be a driver of a vehicle under certain circumstances, it does not define what, or who, the driver of a vehicle is.

There are a number of different situations where a person can be said to be 'driving' a vehicle even though they do not conjure up a picture of conventional driving.

Whether they are 'driving' or not will be a question of fact. This question will be decided by considering the extent to which the person has control of both the direction and movement of the vehicle. The issue of whether or not a person is still driving a vehicle is of particular importance to offences such as driving while disqualified (as to which, **see chapter 11**) and drink driving (**see chapter 5**).

In addition, what the person is actually doing must fall within the ordinary meaning of the word 'driving' (see *McQuaid* v *Anderton* [1980] 3 All ER 540—defendant steering car which was being towed; held to be driving while disqualified).

Other factors which will be considered by the courts will be:

- How long the defendant had control of the movement and direction of the vehicle (*Jones* v *Pratt* [1983] RTR 54—the defendant was a passenger who grabbed the wheel to prevent the driver hitting an animal; held not to be 'driving').
- Whether the defendant set the vehicle in motion deliberately (*Rowan* v *Chief Constable of Merseyside* (1985) The Times, 10 December).
- Whether the defendant used the vehicle's controls in order to direct its movement (*Burgoyne* v *Phillips* [1983] RTR 49).

Applying these principles, pushing a car along while steering it has been held not to be driving so far as the law in England and Wales is concerned (*R* v *MacDonagh* [1974] RTR 372) although in Scotland such behaviour would amount to driving (*Ames* v *MacLeod* 1969 JC 1).

For the purposes of road traffic law, motor cycles are 'driven' as opposed to being 'ridden'.

Straddling a motorcycle and 'pedalling' it along with the feet can amount to driving (*Gunnell* v *DPP* [1994] RTR 151).

Kneeling up on the driver's seat of a car and releasing the handbrake briefly has been held to be 'driving' (see *Rowan* above). For the situation relating to what will amount to 'taking' a motor vehicle under the Theft Act 1968 (**see Crime, chapter 12**).

Recently the Divisional Court has accepted a finding that a person sitting in the driver's seat of a car with the engine running had been 'driving' (*Planton* v *DPP* [2002] RTR 9; [2002] EWHC 450).

Attempting to Drive

Whether a person is 'attempting to drive' is also a question of fact and will usually be determined by the principles applying to attempts generally (**see Crime, chapter 3**).

Simply put, these will include acts which are more than merely preparatory to the act of driving (such as opening a car door). The fact that the vehicle is incapable of being driven (say, because the engine is faulty), will not prevent a charge involving an 'attempt' to drive (*R* v *Farrance* [1978] RTR 225).

If a defendant sits in the driver's seat of a car and, being drunk, tries to put their house keys in the ignition, that behaviour may be enough to prove a charge of 'attempting to drive' while unfit (*Kelly* v *Hogan* [1982] RTR 352 *obiter*). **See also chapter 5**.

In Charge

The principles to be applied when considering whether a person is 'in charge' of a vehicle are set out in *DPP* v *Watkins* [1989] 2 WLR 966. Two different situations might arise:

- Where the defendant is the owner of the vehicle or where they have recently driven it. In these cases it would be for the defendant to show that they were no longer in charge of it and that there was no likelihood of their resuming control at the relevant time (e.g. while they were drunk).
- Where the defendant is not the owner or has not recently driven the vehicle. In these cases the prosecution must show that the defendant was in voluntary control of the vehicle or intended to become so in the immediate future.

In arriving at their decision, courts should consider:

- whether the defendant had the keys for the vehicle;
- where the defendant was in relation to the vehicle at the time;
- what evidence there is of the defendant's intention to take control of the vehicle.

Note that the supervisor of a learner driver can be 'in charge' of the vehicle and can therefore commit an offence if unfit through drink or drugs at the time (*Langman* v *Valentine* [1952] 2 All ER 803) (**see chapter 5**).

Owner

'Owner' in relation to a vehicle subject to a hire purchase agreement is defined under the Road Traffic Act 1988, s. 192. Section 192 provides that the 'owner' in such circumstances means the person in possession of the vehicle under that agreement. Whether a person is the owner of a vehicle in any other case is a question of fact to be determined by the court in each case. Facts including the way in which the alleged owner treats and uses the vehicle, whether they have spent any money on its purchase or upkeep and whether they have taken out insurance in relation to it may all be evidence of 'ownership'.

A 'keeper' is different from an owner but again the question will be one of fact for a court. In any enactment relating to the keeping of vehicles on public roads (e.g. vehicle excise offences, **see chapter 12**), a person so 'keeps' a vehicle if he/she causes it to be on such a road for any period, however short, when it is not in use there (Vehicle Excise and Registration Act 1994, s. 62(2)).

Somewhat confusingly, for the purposes of Part II of the Road Traffic Act 1991 (which regulates parking in London), 'the owner of a vehicle shall be taken to be the person by whom the vehicle is kept' (s. 82(2)).

Whereas an owner probably remains as such until they dispose of the vehicle or pass ownership to someone else, a keeper can cease to be so if they temporarily part with the vehicle (e.g. by leaving it with a garage for repairs) (*R* v *Parking Adjudicator, ex parte The Mayor and Burgesses of the London Borough Council of Wandsworth* (1996) *The Times*, July 22).

Automatism

Where a person's movements are beyond his/her control or his/her actions are brought about involuntarily, he/she will not generally be liable at criminal law as the element of *actus reus* is not present (**see Crime, chapter 2**).

A notable example of this is *automatism*. Automatism occurs where the defendant is temporarily deprived of his/her ability to control his/her movements. When this happens to the driver of a vehicle, it may remove his/her liability for certain offences such as driving without due care. This is because there is no willed action or omission by the defendant. It is also highly unlikely that he/she would have the required state of mind (**see chapter 2**).

Such a situation might be brought about by a swarm of bees flying in through the open window of a moving car (*Hill* v *Baxter* [1958] 1 All ER 193) or the driver lapsing into a coma.

If a defendant is suffering from a particular medical condition or is prone to effects which are likely to impair his/her driving ability (such as dizziness—*R* v *Sibbles* [1959] Crim LR 660), he/she has a duty to take reasonable steps to avoid driving when the symptoms are likely to arise. If a diabetic continues to drive while experiencing the start of a hypoglycaemic episode, it is unlikely that the defence of automatism would be available to a charge of careless driving; the diabetic should have stopped driving until the episode had passed (see *Moses* v *Winder* [1981] RTR 37).

Similarly, where the loss of voluntary movement or control is brought about by self-induced measures (such as taking/failing to take medication or by drinking), automatism is not generally available as a defence (*R* v *Quick* [1973] QB 910).

Driving in a state of reduced awareness brought on by the continuous focusing on a long, featureless road (sometimes called 'motorway hypnosis' is not enough to raise the defence of automatism (*Attorney-General's Reference (No. 2 of 1992)* [1993] RTR 337)).

For a more detailed examination of this and other general defences, **see Crime, chapter 4**. For the defence of 'mechanical defect', **see chapter 2**.

Road

A road is defined under s. 192(1) of the Road Traffic Act 1988 as:

> ... any highway and any other road to which the public has access, and includes bridges over which a road passes,

This definition (or one very similar to it, e.g. under s. 142(1) of the Road Traffic Regulation Act 1984) is applicable to most occasions where the expression 'road' is used in statutes.

Whether the public has access is a question of fact. If only a restricted section of the public (such as members of a club) has access to a road, that is not enough to make it a 'road' (*Blackmore* v *Chief Constable of Devon and Cornwall* (1984) *The Times*, 6 December).

In a recent case the Divisional Court accepted the magistrates' finding that Trafalgar Square was a 'road' for the purpose of ss. 67 and 69 of the Road Traffic Act 1988 (*Sadiku* v *DPP* [2000] RTR 155).

Any access enjoyed by the public must be with the agreement of the landowner. As Lord Sands put it in *Blackmore*, the members of the public must not have obtained access 'either by overcoming a physical obstruction or in defiance of prohibition, express or implied'. Therefore roads are capable of being closed or cordoned off in a way that alters their status as such.

In two recent cases, both involving car parks, the Court of Appeal took a very broad approach to the construction of the definition of a 'road' in relation to the requirements for compulsory insurance (**see chapter 6**). The first case, *Clarke* v *Kato* [1998] 1 WLR 1647, involved a person being struck by an uninsured driver while in the car park behind a parade of shops. The second case, *Cutter* v *Eagle Star Insurance Co. Ltd* (1996) *The Times*, 22

November, involved a passenger being injured while in a vehicle in a multi-storey car park. On appeal to the House of Lords, the plaintiffs in both cases argued that the Road Traffic Act 1988 already made provision for the distinction between roads and car parks and that there was no justification for extending the meaning of the definition of a road beyond its normal meaning. In each of the cases before the House, the car parks were provided solely for the parking of vehicles. The House of Lords accepted the plaintiffs' arguments and overturned the earlier decisions of the Court of Appeal. Whether a particular location falls within the definition under s. 192 of the Road Traffic Act 1988 will ultimately be one of fact for the court to decide, having reference to the character and function of the land in question. Their lordships held that, on that construction, it would only be exceptional circumstances that a car park would be deemed to fall within the definition of a road. Their lordships held that, in the first case, the mere presence of an alleyway which joined onto the car park did not change the ultimate character or purpose of the land while, in the second case, the parking bays within a multi-storey car park could not be regarded as an integral part of the carriageway which led into the car park itself (*Cutter* v *Eagle Star Insurance Co. Ltd, Clarke* v *Kato* [1998] 1 WLR 1647).

In practical terms, the effect of these decisions was limited to offences and regulations which are confined to 'roads'. However, some changes in legislation were needed such as the requirement for compulsory insurance under s. 143(1)(a) of the Road Traffic Act 1988 (**see chapter 6**) and the duty to report accidents under s. 170 of the 1988 Act (**see chapter 4**). Many of the more common or more serious road traffic provisions related to roads *or* 'public places' without the need for amendment (see below) and may still be committed in the sorts of places considered by the House of Lords in the above cases.

Car parks and forecourts may, in exceptional cases, be shown to be roads (see *Baxter* v *Middlesex County Council* [1956] Crim LR 561) but the prosecution would need to show that they fell squarely within the definition under s. 192 in the light of *Clarke* v *Kato* above. As well as showing them to be roads, you can also show some pub car parks to be 'public places'—at least during licensing hours—for the purposes of drink/drive offences (*Sandy* v *Martin* [1974] Crim LR 258).

Generally a road stretches to the boundary fences or grass verges adjacent to it, including any pavements (*Worth* v *Brooks* [1959] Crim LR 885).

If a vehicle is partly on a road and partly on some other privately owned land it can be treated as being 'on a road' for the purposes of road traffic legislation (*Randall* v *Motor Insurers' Bureau* [1969] 1 All ER 21).

'Public roads' which are referred to in the Vehicle Excise and Registration Act 1994 are those roads which are repairable at the public expense. Clearly this is a much more restrictive definition than that under the Road Traffic Act 1988.

Public Place

In order to prove that a place is in fact a 'public place' for the purposes of road traffic offences, it must be shown by the prosecution that:

- those people who are admitted to the place in question are members of the public and are admitted as such, not as members of some special or particular *class* of the public (e.g. people belonging to an exclusive club) or as a result of some special characteristic that is not shared by the public at large; *and*
- those people are so admitted with the permission, express or implied, of the owner of the land in question

(*DPP* v *Vivier* [1991] RTR 205).

Whether a place is in fact a 'public place' will be a question of fact for the court. This means that different magistrates' courts can arrive at different decisions when presented with the same facts. Some examples of places which have been held to qualify as 'public places' are:

- a privately-owned caravan site open to campers (*DPP* v *Vivier* above);
- a school playground used outside school hours as a leisure park by members of the public (*Rodger* v *Normand* [1995] SLT 411);
- the 'Inward Freight Immigration Lanes' at Dover Eastern Docks (*DPP* v *Coulman* [1993] RTR 230);
- a field used in connection with an agricultural show (*Paterson* v *Ogilvy* 1957 SLT 354);
- a multi-storey car park (*Bowman* v *DPP* [1991] RTR 263).

A recent case in Essex involving a causeway between the mainland and an island provides a good illustration of the process that a court will go through in arriving at its decision in this area. In this case—which involved a motorist sitting in his car with the engine running on the causeway—the Divisional Court held that:

- the use of the causeway by members of the public was extremely limited;
- those visitors were limited to a 'special class of public' as discussed in *DPP* v *Vivier* (above);
- no invitation or permission was given to the general public and there was no evidence of general public access to the island;
- the only real access applied to residents and people with purposes connected to those residents.

Therefore, on these facts, the causeway was held not to be a 'public place' within the meaning of s. 5(1)(a) of the Road Traffic Act 1988 (e.g. for the purposes of drink driving—**see chapter 5**)—*Planton* v *DPP* [2002] RTR 9. The court accepted however that, by sitting in the car with the engine running, the defendant had been 'driving' (see above).

These requirements should not be confused with the test for whether a piece of land amounts to 'a road' (see above). The importance of police officers providing enough evidence to show that a particular location amounted to a public place was highlighted in *R* v *DPP, ex parte Taussik* [2001] ACD 10. In that case the defendant was stopped as she drove out of an access road leading from a block of flats. The road was a cul-de-sac leading off a highway and was maintained by the local housing department. At the entrance to it there was large sign saying 'Private Residents Only'. Following her conviction under s. 5(1)(b) of the Road Traffic Act 1988 (as to which, **see chapter 5**), the defendant appealed, alleging that the sign on the access road excluded anyone other than residents from using it and therefore was not a 'public place'. The Divisional Court took the view that the evidence of the police officers as to the actual use of the road by other people was very thin. As there was no evidence from the officers themselves that they had seen motorists (other than residents) using the road, the court was unable to conclude that the road was anything other than a private one. The court held that, as the issue of whether a place is a 'public place' or not is largely a question of fact, it is essential that the prosecution present clear evidence showing who uses the road, when and for what purpose.

For a place which is ostensibly *public* in its nature to become a *private* place, either permanently or temporarily, there needs to be some form of physical obstruction to be overcome in order to enter that place (*R* v *Waters* (1963) 107 SJ 275). Therefore a pub landlord, by ordering people to leave the car park of his pub, had not done enough to turn what was a public place (the car park during opening hours) into a private place.

In deciding whether or not a place is in fact a 'public place', magistrates may use their own local knowledge, but must be circumspect in doing so (*Clift* v *Long* [1961] Crim LR 121).

Highway

A 'highway' is a way over which the public has a right to pass and re-pass by foot, horse or vehicle, or with animals (*Lang* v *Hindhaugh* [1986] RTR 271).

For a highway to exist, there must be some form of 'dedication' of the relevant land to the public and, once so dedicated, it is unlikely that the public status of a highway can be changed.

Unlike a road, a highway does not cease to be such when it is temporarily roped off or closed (*McCrone* v *and J and L Rigby (Wigan) Ltd* (1950) 50 LGR 115).

Highways will include public bridleways and footways; they also include public bridges over which they pass. The further definitions of bridleways, footways and carriageways are not of general relevance. These definitions are set out in s. 329 of the Highways Act 1980 and are only relevant to a small number of specific offences (**see chapter 8**). Broadly, footways are the bits of a highway that you walk on and carriageways are the bits that you drive on. Bridleways, as the name suggests, are highways where the public have a right to ride or lead horses (and related animals) or to pass on foot. Some of these definitions have been reclassified by the Countryside and Rights of Way Act 2000 which creates a number of different types of byway with differing rights of access on foot, horseback or with vehicles that are not mechanically propelled. These definitions are relevant when considering issues of obstruction and off-road driving (**see chapter 8**).

The relevant local authority has an *absolute* duty to maintain the fabric of the highway (i.e. the road surface) under s. 41 of the 1980 Act. However, that absolute duty does not extend to keeping the highway free from ice and snow (see *Goodes* v *East Sussex County Council* [2000] 1 WLR 1356).

The Greater London Authority Act 1999 makes provision for certain highways to be designated as 'GLA roads' (s. 260).

Use, Cause or Permit?

The concepts of using, causing or permitting are central to many road traffic offences, particularly in relation to construction and use (**see chapter 9**). They are also relevant in the fixed penalty system which is not available for offences of 'causing or permitting' (**see chapter 14**).

Use

One of the difficulties with 'using' as a concept is that it is contextual. In deciding the many cases that have grown up around the word 'use', the courts have often interpreted the expression in a way that is specific to its statutory context. This means that there is no hard and fast 'definition' that can be applied to every circumstance. 'Using' a motor vehicle in relation to an insurance offence will be interpreted differently from the same expression in a construction and use offence.

'Using' has been held to involve an element of 'controlling, managing or operating the vehicle as a vehicle' (*Thomas* v *Hooper* [1986] RTR 1).

The 'using' of a vehicle is generally, though not exclusively, restricted to:

- the driver;
- the driver's employer (when the driver is driving the vehicle on the employer's business).

In the *West Yorkshire Trading Standards* case (below) the Court of Appeal restricted the meaning of the word 'use' when it appears alongside 'cause' and 'permit' (see below). A good example of this can be found in the numerous construction and use offences (as to which, **see chapter 9**).

In such circumstances 'using' is restricted to the driver or owner of the vehicle and, in the case of the owner, only if he/she employed the driver under a contract of service and at the material time the vehicle was being driven on the owner's business. Where the expression 'use' does not appear alongside 'cause' and 'permit' the concept is wider. Practically, if there is any doubt as to the appropriate wording of a charge/summons, the advice of the Crown Prosecution Service should be sought.

If a person driving a vehicle is doing so in the ordinary course of his/her employer's business, the employer is *using* the vehicle (see e.g. *West Yorkshire Trading Standards Service* v *Lex Vehicle Leasing Ltd* [1996] RTR 70). In such a case you must prove that:

- the defendant (employer) actually owned the vehicle;
- at the relevant time, the driver was employed by the defendant; and
- the driver was driving the vehicle in the ordinary course of his/her employment

(*Jones* v *DPP* [1999] RTR 1).

It is immaterial that the employer has not specifically authorised the employee to use the vehicle in such a way (*Richardson* v *Baker* [1976] RTR 56).

If the driver is not an employee then the employer is not using the vehicle, even if that driver is a partner of the firm or has been asked to drive the vehicle by the employer (*Crawford* v *Haughton* [1972] 1 WLR 572). The employer may, however, be shown to be 'causing' or 'permitting' (see below).

If a garage lends one of its vehicles to a customer and the customer then uses the vehicle on his/her own business, the garage cannot be said to be 'using' the vehicle (see *Dove (LF)* v *Tarvin* (1964) 108 SJ 404).

If a vehicle is shown to be a 'motor vehicle' (**see para. 1.2.1**) and on a 'road' (**see para. 1.2.7**), it may be said to be in 'use' even if it is in such a state that it cannot be driven (*Pumbien* v *Vines* [1996] RTR 37). There is now no longer a need to show some element of control or operation of the vehicle by the owner in order to prove 'use'.

The *intention* of the owner is irrelevant in determining whether or not a vehicle was being 'used' at the time of an offence; what matters is that the defendant can be shown to have 'had the use of' that vehicle (*Eden* v *Mitchell* [1975] RTR 425).

Therefore, if a vehicle is not in a roadworthy condition it can still be available for the owner's use even though the owner has no intention of utilising it while it remains in that state.

For the owner of a vehicle to be convicted of 'using' a vehicle without insurance when it is being driven by someone else, you must prove that the defendant was in fact the owner *and* that the driver was his/her employee acting in the course of their employment (*Jones* v *DPP* [1999] RTR 1).

It should be noted that the Divisional Court has applied a broader interpretation to the meaning of 'use' when considering offences involving *trailers*. The court held that the owner of a defective trailer who is responsible for putting it on a road should not be able to escape liability for its condition simply by arguing that it was being drawn—and therefore 'used'—by someone else (*NFC Forwarding Ltd* v *DPP* [1989] RTR 239).

Passengers as Users

A pillion passenger on a motor cycle or a passenger in a car does not generally 'use' the vehicle (see e.g. *Hatton* v *Hall* [1997] RTR 212). However, if a passenger arranges to travel in or on a vehicle for his/her own benefit, he/she will 'use' the vehicle (*Cobb* v *Williams* [1973] RTR 113). A passenger can also 'use' the vehicle if there is an element of 'joint enterprise' (*O'Mahoney* v *Joliffe and Motor Insurers' Bureau* [1999] RTR 245; see also **Crime, chapter 2**).

Cause

'Causing' will involve some degree of 'control' or 'dominance' by, or some express mandate from, the *causer*.

Causing requires both positive action and knowledge by the defendant (*Price* v *Cromack* [1975] 1 WLR 988). If the owner of a vehicle is to be shown to have 'caused' an offence to be committed, he/she must be shown to have done something to contribute to it (e.g. by instructing another to drive it) *and also* to have known of any relevant facts (e.g. that it was overloaded) (*Ross Hillman Ltd* v *Bond* [1974] QB 435).

In cases involving employers' vehicles it may be easier to prove 'use' of the vehicle (see above).

Wilful blindness by employers to their employees' unlawful actions (e.g. not completing drivers' records) is not enough to amount to 'causing' the offence (*Redhead Freight Ltd* v *Shulman* [1989] RTR 1).

Permit

Permitting is less direct or explicit than causing. Permitting involves giving leave or license to do something (see *Houston* v *Buchanan* [1940] 2 All ER 179). The relevant permission (e.g. to use a vehicle in a certain way or subject to some proviso) can be express or it can be inferred by the relevant person (such as where someone is given the use of a friend's car without any conditions being stipulated by the friend).

Generally, in order to prove an offence of 'permitting', you will need to show knowledge by the defendant of the vehicle's *use*, and the *unlawful nature* of that use. However, it is necessary to consider the relevant piece of legislation in each case, together with its intended purpose. (For the situation regarding insurance, **see chapter 6**.)

You cannot 'permit' yourself to do something (*Keene* v *Muncaster* [1980] RTR 377).

If there is any doubt as to which offence is appropriate it is acceptable to charge 'using' *or* 'causing' *or* 'permitting' as alternatives (see *Ross Hillman Ltd* above) but the advice of the local Crown Prosecution Service should be sought.

2 Offences involving standards of driving

2.1 Introduction

This chapter deals with the main offences involving driving below the required standard or in a way which presents a danger to others.

Those offences which involve drink or drugs are covered in **chapter 5**, although the offence of causing death by careless driving while over the prescribed limit is dealt with here.

Reference should also be made, where appropriate, to the CPS charging standards, **see appendix 4**.

When considering the various definitions within the offences (e.g. 'drive', 'motor vehicle'), you should refer to **chapter 1**.

2.2 Causing Death by Dangerous Driving

OFFENCE: **Causing Death by Dangerous Driving—*Road Traffic Act 1988, s. 1***

- Triable on indictment • Ten years' imprisonment
- Obligatory disqualification—minimum two years • Compulsory re-test.

(Serious arrestable offence)

The Road Traffic Act 1988, s. 1 states:

> A person who causes the death of another person by driving a mechanically propelled vehicle dangerously on a road or other public place is guilty of an offence.

KEYNOTE

This offence applies to a 'mechanically propelled vehicle'—a wider term than 'motor vehicle' (**see chapter 1**) and one which includes dumper trucks, cranes, trials and quad bikes.

It can be committed on a road or in other public places which again give it a wider effect.

Provided the basic elements (mechanically propelled vehicle on a road/public place) are met, you must prove that:

- the defendant caused the death of another person
- the defendant drove dangerously.

Where the defendant is charged under s. 1, evidence of drink will be admissible where the quantity of it may have adversely affected the quality of the his/her driving (*R* v *Woodward* [1995] RTR 130). For a full discussion of the specific 'drink drive' offences, **see chapter 5**. For the offence of causing death by careless driving when under the influence of alcohol, **see para. 2.7**.

Section 36 of the Road Traffic Offenders Act 1988 (requiring the court to disqualify a person convicted of certain offences until they have passed the relevant test) applies to this offence (**see para. 11.5.1**). This means that, if convicted of this offence, the defendant will have to take an extended driving test before he/she can get his/her licence back.

2.2.1 Causing Death of Another

The death must be that of a person other than the defendant. This has been held (in a Scottish case) to include a foetus which was later born alive but which subsequently died (*McCluskey* v *HM Advocate* [1989] SLT 175).

In deciding an extradition case, the House of Lords held that the offence under s. 1 had not repealed the common law offence of manslaughter and, as such, a driver causing the death of another in this way could still be indicted for homicide (*R* v *Governor of Holloway Prison, ex parte Jennings* [1983] 1AC 624 and **see Crime, chapter 5**). However, a charge of manslaughter will rarely be brought in such cases. There is some concern that the maximum sentence available to a judge in the event of a conviction for a s. 1 offence (ten years' imprisonment) is not enough to reflect the feelings of society or individual victims.

The driving by the defendant must be shown to have been *a* cause of the death; it is not necessary to show that it was the sole or even a substantial cause of death (*R* v *Hennigan* [1971] 3 All ER 133). Therefore it is irrelevant whether or not the person killed contributed to the incident which resulted in his/her death.

2.2.2 Dangerous Driving

Section 2A of the Road Traffic Act 1988 states:

(1) For the purposes of sections 1 and 2 above a person is to be regarded as driving dangerously if (and, subject to subsection (2) below, only if)—

(a) the way he drives falls far below what would be expected of a competent and careful driver, and

(b) it would be obvious to a competent and careful driver that driving in that way would be dangerous.

(2) A person is also to be regarded as driving dangerously for the purposes of sections 1 and 2 above if it would be obvious to a competent and careful driver that driving the vehicle in its current state would be dangerous.

(3) In subsections (1) and (2) above 'dangerous' refers to danger either of injury to any person or of serious damage to property; and in determining for the purposes of those subsections what would be expected of, or obvious to, a competent and careful driver in a particular case, regard shall be had not only to the circumstances of which he could be expected to be aware but also to any circumstances shown to have been within the knowledge of the accused.

(4) In determining for the purposes of subsection (2) above the state of a vehicle, regard may be had to anything attached to or carried on or in it and to the manner in which it is attached or carried.

KEYNOTE

This is clearly an *objective* test which focuses, at s. 2A(1)(a) on the manner of driving rather than the defendant's state of mind and, at (b) on what would have been obvious to a hypothetical 'competent and careful driver'. An example is where a diabetic driver drives on a road in the knowledge that he/she is likely to suffer a hypoglycaemic episode (*R* v *Marison* [1997] RTR 457).

In *Attorney-General's Reference (No. 4 of 2000)* [2001] RTR 415, the Court of Appeal reviewed the requirements of s. 2A. That case involved a bus driver who had inadvertently pressed the accelerator pedal instead of the brake, killing two pedestrians. The Court held that, under s. 2A the test is an objective one and there is no requirement to show any specific intent to drive dangerously. It is for the jury to determine what constituted dangerous driving. The Court held that the relevant *actus reus* (**see Crime, chapter 2**) is the act of driving in a manner which was either dangerous (in the case of a dangerous driving charge) or without due care and attention in the case of the alternative offence. Where, as in this case, the driver had been conscious of the act he was performing, it was no defence to claim that he had not intended to press the accelerator. That was more a matter for mitigation than guilt.

An example of the application of s. 2A(2) can be seen in a Scottish case where the defendant drove a small go-kart on a busy road. The nature of the kart and the likelihood of its not being seen by other motorists were held to be features which could be taken into account in assessing whether or not the defendant's driving was in fact 'dangerous' (*Carstairs* v *Hamilton* 1998 SLT 220).

The standard of driving must be shown to have fallen far below that expected of a competent and careful driver; minor driver errors would not amount to such behaviour (but contrast offences under s. 3 below).

In determining what would have been obvious to a competent and careful driver, ss. 2A(3) and (4) introduce a *subjective* element by taking account of circumstances known to the defendant. This unusual mixture of tests means that, although a defendant's behaviour will be judged against the ordinary standards of competent and careful drivers, the defendant's conduct will also be assessed in the light of facts personally known to him/her (such as knowledge of the risk of a load falling off the vehicle (*R* v *Crossman* [1986] RTR 49)). If the vehicle involved was in a dangerous condition it is important that you prove either:

- that the dangerous condition would itself have been obvious to a competent or careful driver; or
- that the defendant actually *knew* of its dangerous condition

(see *R* v *Strong* [1995] Crim LR 428).

The defendant's belief, however honestly held, as to the conditions surrounding his/her driving at the time are not relevant to the issue of whether he/she drove competently and carefully (see *R* v *Collins* [1997] RTR 439). In *Collins* a police driver went through a red traffic light at almost 100 mph colliding with another vehicle and killing two people. His belief that the traffic at the lights was being operated by other officers was not a relevant factor for the jury in considering whether or not his driving had been dangerous. (For police drivers generally, **see para. 2.10.**)

2.3 Dangerous Driving

OFFENCE: **Dangerous Driving—*Road Traffic Act 1988, s. 2***

- Triable either way • Two years' imprisonment and/or a fine on indictment; six months' imprisonment and/or statutory maximum summarily
- Obligatory disqualification • Compulsory re-test

(No specific power of arrest)

The Road Traffic Act 1988, s. 2 states:

> A person who drives a mechanically propelled vehicle dangerously on a road or other public place is guilty of an offence.

KEYNOTE

The elements of this offence are the same as those for s. 1.

Evidence showing how the particular vehicle was being driven before the incident itself may be given in support of the charge of dangerous driving. Where the dangerous driving leads to an accident, the court may allow a police officer who is an expert in the investigation of accidents to give evidence of opinion as to the cause of that accident (*R* v *Oakley* [1979] RTR 417).

In addition, there is a rarely used offence of causing bodily harm by wanton or furious driving or racing (Offences Against the Person Act 1861, s. 35). The offence applies to any vehicle or carriage and is not restricted to roads.

Section 36 of the Road Traffic Offenders Act 1988 (requiring the court to disqualify a person convicted of certain offences until they have passed the relevant test) applies to this offence (**see para. 11.5.1**). This means that, if convicted of this offence, the defendant will have to take an extended driving test before he/she can get his/her licence back.

2.4 Careless and Inconsiderate Driving

OFFENCE: **Careless and Inconsiderate Driving—*Road Traffic Act 1988, s. 3***

- Triable summarily • Fine • Discretionary disqualification

(No specific power of arrest)

The Road Traffic Act 1988, s. 3 states:

> If a person drives a mechanically propelled vehicle on a road or other public place without due care and attention, or without reasonable consideration for other persons using the road or place, he is guilty of an offence.

KEYNOTE

Like the offences under ss. 1 and 2, this offence also applies to a mechanically propelled vehicle and to public places as well as roads.

Where a constable in uniform has reasonable grounds for believing that a mechanically propelled vehicle is being used or has been used on any occasion in a manner which contravenes ss. 3 or 34 (off-road driving—see chapter 8) and is causing (or is likely to cause), alarm, distress or annoyance to members of the public, he/she has the powers set out in the Police Reform Act 2002, s. 59 (as to which, see para. 2.4.1).

'Due care and attention' is the standard of driving that would be expected of a reasonable, prudent and competent driver in all the attendant circumstances.

There is one *objective* standard of driving which is expected of all drivers—even learner drivers (*McCrone* v *Riding* [1938] 1 All ER 157). Once you have proved that a defendant departed from that standard of driving, and that his/her actions were 'voluntary', then the offence is complete. There is no need to prove any *knowledge* or *awareness* by the defendant that their driving fell below that standard (*R* v *Lawrence* [1981] RTR 217).

The standard of driving that would be expected of a prudent and competent driver will be a question of fact for the court to decide and, in so deciding, the magistrate(s) may take into account local factors such as the expected level of traffic, the time of day, peculiar hazards, etc. (*Walker* v *Tolhurst* [1976] RTR 513).

There is a considerable overlap between the criminal and civil law in the area of careless driving, mainly because of the relevant civil duty of care owed by drivers to other road users and also because claims for damages against drivers are involved in many accidents and collisions. For this reason many of the authorities cited in identifying the principles of 'due care and attention' are civil ones.

Evidence of earlier incidents involving careless or inconsiderate driving around the same time as the offence charged may be admissible to support that charge under some circumstances (*Hallett* v *Warren* (1926) 93 JP 225) as it may if the offence is charged as one continuing offence (*Horrix* v *Malam* [1984] RTR 112).

If a witness reports the driver of an unidentified vehicle as having committed a driving offence, it is critical that the witness provides direct evidence of what they actually saw. It would not be enough to produce evidence from the police to show that they had received details of the vehicle which they subsequently traced back to the driver. Both vehicle *and* driver must be linked by admissible and relevant evidence (*Ahmed* v *DPP* [1998] RTR 90).

If a driver falls asleep at the wheel he/she will be guilty of careless driving (*Henderson* v *Jones* (1955) 119 JP 304) but evidence of this fact from the driver alone will not be enough to support a charge under s. 3 (*Edwards* v *Clarke* (1951) 115 JPN 426). In order to prove that a driver drove inconsiderately you must show that some other person using the road was actually inconvenienced (*Dilks* v *Bowman-Shaw* [1981] RTR 4).

Other persons using the road/public place can include pedestrians who are deliberately sprayed with water from a puddle or passengers in a vehicle (see *Pawley* v *Wharldall* [1965] 2 All ER 757).

If a motorist takes action in response to an emergency situation, his/her actions are to be judged against what was a 'reasonable' course of action in those circumstances in assessing whether or not the driving amounted to an offence (*Jones* v *Crown Court at Bristol* [1986] RTR 259).

Breaching certain road traffic regulations will always be potentially relevant evidence of poor driving but will not always be conclusive of the issue. For instance, although colliding with another vehicle has been held not to amount to sufficient evidence in itself of careless driving, crossing a central white line without explanation has (*Mundi* v *Warwickshire Police* [2001] EWHC Admin 448). (See also *Bensley* v *Smith* [1972] Crim LR 239.) However, simply breaching the regulations at a pedestrian crossing (**see chapter 8**) is not of itself proof that the person's driving fell below the required standard (*Gibbons* v *Kahl* [1956] 1 QB 59). On the other hand, just because a traffic signal is showing a green light does not mean that a driver is entitled to assume that no other person or vehicle might be proceeding from another direction and it may be that, in the circumstances, a reasonably careful driver would have anticipated that a pedestrian or other road user might still move into his/her path (see *Goddard and Walker* v *Greenwood* [2003] RTR 10). Although police and other emergency service drivers enjoy some limited exemption from traffic regulation (**see para. 2.10**). they will be subject to the general rules in relation to the standard of driving expected of motorists. Simply using the relevant audible and visual warning equipment fitted to a police vehicle will not absolve the driver from exercising due care and attention towards other road users.

2.4.1 Vehicles Used for Causing Harassment etc.

Section 59 of the Police Reform Act 2002 states:

(1) Where a constable in uniform has reasonable grounds for believing that a motor vehicle is being used on any occasion in a manner which—
 (a) contravenes section 3 or 34 of the Road Traffic Act 1988 (c. 52) (careless and inconsiderate driving and prohibition of off-road driving), and
 (b) is causing, or is likely to cause, alarm, distress or annoyance to members of the public,
 he shall have the powers set out in subsection (3).

(2) A constable in uniform shall also have the powers set out in subsection (3) where he has reasonable grounds for believing that a motor vehicle has been used on any occasion in a manner falling within subsection (1).

KEYNOTE

This legislation gives uniformed police officers additional powers to deal with motor vehicles being used in the unlawful, anti-social or just plain annoying ways described. For the two specific offences described, **see para. 2.4 and chapter 8**.

A peculiar and complicating feature is the definition of a motor vehicle here. For the purposes of these powers above, 'motor vehicle' means any mechanically propelled vehicle *whether or not it is intended or adapted for use on roads* (s. 59(9)). While making practical sense (given that one of the main circumstances in which the powers will be used is dealing with off-road driving), this definition conflicts with the same expression when used under s. 185 of the Road Traffic Act 1988 which is applied by most road traffic legislation (**see chapter 1**). The definition of motor vehicle here will cover everything from go-karts and home-made trials bikes to dumper trucks and building site vehicles.

The requirement for the constable to have 'reasonable grounds for believing' in the facts set out above is greater than mere suspicion or even belief (**see General Police Duties, chapter 2**). He/she must be able to point to the existence of *reasonable grounds* giving rise to a belief that the motor vehicle is being or has been used in one of the ways described.

The powers available in these circumstances are listed in s. 59(3):

(a) power, if the motor vehicle is moving, to order the person driving it to stop the vehicle;

(b) power to seize and remove the motor vehicle;

(c) power, for the purposes of exercising a power falling within paragraph (a) or (b), to enter any premises on which he has reasonable grounds for believing the motor vehicle to be;

(d) power to use reasonable force, if necessary, in the exercise of any power conferred by any of paragraphs . . . (a) to (c).

Although the power of entry excludes entry into a 'private dwelling house' (see s. 59(7)), that definition does not include garages or other structures occupied with the dwelling, nor driveways or other appurtenant land (see s. 59(9)). This definition of private dwelling is therefore more restricted than similar expressions under, say, the Police and Criminal Evidence Act 1984 (**see General Police Duties, chapter 2**) and means that the above power is available in relation to garages and driveways of houses.

The powers under s. 59(3) above are among those that can be conferred on a Community Support Officer (CSO) designated under sch. 4 to the Police Reform Act 2002 (**see General Police Duties, chapter 2**). However, a designated CSO cannot enter any premises in the exercise of those powers unless he/she is in the company *and* under the supervision of a constable (sch. 4, para. 9(2)).

A vehicle cannot be seized under the s. 59 power unless the officer:

- has warned the person appearing to the officer to be the person whose use falls within subsection (1) that he/she will seize it, if that use continues or is repeated; and
- it appears to the officer that the use has continued or been repeated after the warning

(s. 59(4)).

However, a warning is not required if:

- the circumstances make it impracticable for the officer to give a warning;
- the officer has already on that occasion given such a warning in respect of any use of that motor vehicle or of another motor vehicle by that person or any other person;
- the officer has reasonable grounds for believing that such a warning has been given on that occasion otherwise than by him/her; or
- the officer has reasonable grounds for believing that the person whose use of that motor vehicle on that occasion would justify the seizure is a person to whom a warning under that subsection has been given (whether or not by that officer or in respect the same vehicle or the same or a similar use) on a previous occasion in the previous 12 months

(s. 59(5)).

Reference to warning the person 'using' the vehicle here makes the requirements broader than simply warning the relevant *driver*. For a discussion of occasions where a vehicle may be being 'used' by someone other than the driver, **see chapter 1**.

The above conditions where a warning will not be needed are alternatives and any one of them will suffice. The sensible provision that a warning is not needed where it would be impracticable to give one is also fairly wide and does not mean the officer has to show it was *impossible* to do so, or even very difficult.

Given the provisions in relation to earlier warnings that may have been given by that or other officers, it will be important from a practical point of view to maintain accurate records of any warnings given. Because the exercise of the power of seizure amounts to an interference with property, careful consideration should be given to the Human Rights Act 1998, and in particular Protocol 1, Article 1 of the European Convention on Human Rights (**see General Police Duties, chapter 2**).

For the relevant arrangements dealing with the removal, retention, release and disposal of vehicles seized under this power, see the Police (Retention and Disposal of Motor Vehicles) Regulations 2002 (SI 2002 No. 3049).

OFFENCE: **Failing to Stop—*Police Reform Act 2002, s. 59(6)***

- Triable summarily • Fine

(No specific powers of arrest)

The Police Reform Act 2002, s. 59 states:

> (6) A person who fails to comply with an order under subsection (3)(a) is guilty of an offence...

KEYNOTE

The power under s. 59(3)(a) is to order the person *driving* the motor vehicle in the relevant circumstances to stop. 'Driving' has the same meaning as in the Road Traffic Act 1988 (**see chapter 1**). If you are going to prosecute someone for failing to comply with the order, you will first need to prove that the order was properly given (e.g. that the person giving it had the authority to do so under the circumstances) and that the order was both heard and understood. The first of those matters will be of particular importance to Community Support Officers. For the general power to stop vehicles, **see chapter 10**. In the context of the general power to stop vehicles on roads, 'stop' has been held to mean bringing the vehicle to a halt and remaining at rest for long enough for the officer to exercise whatever additional powers are appropriate (see *Lodwick* v *Sanders* [1985] 1 All ER 577). If the same interpretation is given to the power here, 'stopping' must be at least long enough to allow the officer to deliver the statutory warning if appropriate, or perhaps to check whether any such warning has already been given. Although the power does not extend to demanding the keys from the driver, the section clearly gives uniformed officers powers of enforcement, including a power to use reasonable force if necessary.

Interestingly, there is no requirement that any failure to comply be without reasonable excuse or anything of that nature. 'Fail' ought to include refusal, although the absence of any express wording to that effect (as found in the drink driving legislation, **see chapter 5**) will no doubt be raised on behalf of some deserving defendant.

2.4.2 National Driver Improvement Scheme

Within the North Report which led to the changes brought about by the Road Traffic Act 1991, was a recognition that it is in the public interest to retrain some traffic offenders rather than to punish them. As a result, a Driver Rectification Scheme was introduced in the Devon and Cornwall Constabulary area for some drivers who had come to police attention as a result of reported offences of driving without due care and attention. The scheme was quickly adopted by other police forces and, with the support of the HMIC and the government, the scheme developed into the National Driver Improvement Scheme. Most forces in England and Wales now use the scheme by which the police offer a course of re-training to drivers who meet the relevant criteria as an alternative to prosecution.

2.5 Aiding and Abetting

It is possible for someone to aid and abet another to commit an offence under ss. 1–3 of the Road Traffic Act 1988. The supervisor of a learner driver may commit such an offence if he/she fails to supervise the other person properly (*Rubie* v *Faulkner* [1940] 1 All ER 285). (For the supervision of learner drivers generally, **see chapter 11**.) A driving test examiner however is in the vehicle for a different purpose and will not generally be liable for the driver's behaviour (*British School of Motoring Ltd* v *Simms* [1971] 1 All ER 317).

However, although driving test examiners are not 'supervisors' or instructors, there are occasions where the driving of the examinee may be so incompetent that it would be a danger to other road users to allow it to continue. As such, the examiner may be liable as an accessory to any relevant offence. Support for this view can be found in an unreported appeal to the Crown Court at Burnley (*R* v *Charles Hoyle*, 1980) where the judge upheld the examiner's conviction of aiding and abetting the offence of driving without due care and attention.

2.6 Dangerous, Careless and Inconsiderate Cycling

There are similar offences which regulate the standard of cycling *on roads*. These are found under ss. 28 and 29 of the Road Traffic Act 1988 and are summary offences punishable with a fine. The offence of causing bodily injury by 'furious driving' also applies to pedal cycles (**see para. 2.3**). For other offences involving pedal cycles, **see chapter 15**.

'Cycles' will include tricycles and any cycle having four or more wheels (s. 192 of the 1988 Act).

2.7 Causing Death by Careless Driving when under the Influence of Drink or Drugs

OFFENCE: **Causing Death by Careless Driving when under the Influence of Drink or Drugs—*Road Traffic Act 1988, s. 3A***

- Triable on indictment • Ten years' imprisonment
- Obligatory disqualification—minimum two years

(Serious arrestable offence)

The Road Traffic Act 1988, s. 3A states:

(1) If a person causes the death of another person by driving a mechanically propelled vehicle on a road or other public place without due care and attention, or without reasonable consideration for other persons using the road or place, and—
   (a) he is, at the time when he is driving, unfit to drive through drink or drugs, or
   (b) he has consumed so much alcohol that the proportion of it in his breath, blood or urine at that time exceeds the prescribed limit, or
   (c) he is, within 18 hours after that time, required to provide a specimen in pursuance of section 7 of this Act, but without reasonable excuse fails to provide it,

   he is guilty of an offence.

(2) For the purposes of this section a person shall be taken to be unfit to drive at any time when his ability to drive properly is impaired.

(3) Subsection (1)(b) and (c) above shall not apply in relation to a person driving a mechanically propelled vehicle other than a motor vehicle.

KEYNOTE

The elements in relation to the 'causing' of death and 'another person' are the same as those under s. 1 (**see para. 2.2.1**).

The elements relating to 'due care and attention' and 'reasonable consideration' are the same as those under s. 3 (**see para. 2.4**).

The elements relating to 'unfitness to drive' and the consumption of alcohol should be read in conjunction with those discussed in **chapter 5**.

Although the offence under s. 3A applies to a road or other public place, note that s. 3A(1)(b) and (c) can only be committed by a driver of a *motor vehicle* (**see para. 1.2.1**).

The requirement under s. 3A(1)(c) is for the provision of a specimen for *analysis* under s. 7 of the Road Traffic Act 1988 and not a screening breath test under s. 6.

It would appear that the request must be made within 18 hours *after the driving which caused the death* and not after the death itself.

Section 36 of the Road Traffic Offenders Act 1988 (requiring the court to disqualify a person convicted of certain offences until they have passed the relevant test) applies to this offence (**see para. 11.5.1**). This means that, if convicted of this offence, the defendant will have to take an extended driving test before he/she can get his/her licence back.

2.8 The Highway Code

One way of illustrating the standard of driving expected of a prudent and competent driver is through the Highway Code. Section 38 of the Road Traffic Act 1988 states:

> (7) A failure on the part of a person to observe a provision of the Highway Code shall not of itself render that person liable to criminal proceedings of any kind but any such failure may in any proceedings (whether civil or criminal, and including proceedings for an offence under the Traffic Acts, the Public Passenger Vehicles Act 1981 or sections 18 to 23 of the Transport Act 1985) be relied upon by any party to the proceedings as tending to establish or negative any liability which is in question in those proceedings.

Among its many provisions, the Highway Code makes reference to the use of mobile phones and in-car communications equipment (rule 43) and this provision has supported a charge under s. 3 in a Scottish case (*Rae* v *Friel* 1993 SLT 791).

The braking distances shown in the Highway Code however are not admissible in proving speeding cases as they amount to hearsay (*R* v *Chadwick* [1975] Crim LR 105; for hearsay generally, **see Evidence and Procedure, chapter 11**).

The use of mobile phones has caused a lot of excitement in driving debates and has led to convictions for driving without due care and attention (**see para. 2.4**). At the time of writing, the government had concluded its further consultations and more proposals are awaited.

2.9 Identity of Drivers

Section 172 of the Road Traffic Act 1988 states:

> (1) This section applies—
> (a) to any offence under the preceding provisions of this Act except—
> (i) an offence under Part V, or
> (ii) an offence under section 13, 16, 51(2), 61(4), 67(9), 68(4), 96 or 120, and to an offence under section 178 of this Act,
> (b) to any offence under sections 25, 26 or 27 of the Road Traffic Offenders Act 1988,
> (c) to any offence against any other enactment relating to the use of vehicles on roads, except an offence under paragraph 8 of Schedule 1 to the Road Traffic (Driver Licensing and Information Systems) Act 1989, and
> (d) to manslaughter, or in Scotland culpable homicide, by the driver of a motor vehicle.
>
> (2) Where the driver of a vehicle is alleged to be guilty of an offence to which this section applies—
> (a) the person keeping the vehicle shall give such information as to the identity of the driver as he may be required to give by or on behalf of a chief officer of police, and
> (b) any other person shall if required as stated above give any information which it is in his power to give and may lead to identification of the driver.
>
> (3) Subject to the following provisions, a person who fails to comply with a requirement under subsection (2) above shall be guilty of an offence.

KEYNOTE

Section 172 applies to offences other than those listed at s. 172(1)(a)–(d). The excepted offences listed at s. 172(1)(a) and (b) are offences relating to:

- driving instructors
- motoring events on public highways
- protective headgear

- testing of goods vehicles
- regulations for 'type approval'
- obstruction of a vehicle examiner
- requirements to proceed to a place of vehicle inspection
- uncorrected eyesight
- regulations for licensing of large goods vehicle/passenger carrying vehicle drivers
- unlawful vehicle taking in Scotland, and
- post-conviction offences under the Road Traffic Offenders Act 1988.

Therefore s. 172 will apply to regulations made under the applicable sections (e.g. the Road Vehicles (Construction and Use) Regulations 1986; **see chapter 9**).

The requirement under s. 172 applies to the person *keeping* the vehicle and will apply to a person who is the keeper of the vehicle at the time the requirement is made even if he/she was not the keeper at the time of the alleged offence (*Hateley* v *Greenough* [1962] Crim LR 329). The second part of the power (s. 172(2)(b)) applies to *any other person*. Failing to give the information is a summary offence punishable by a fine and, in some circumstances, carrying a power for discretionary disqualification (**see appendix 3**).

There is no particular form of words to be used when making the requirement. It must be shown that the person making the requirement did so by, or on behalf of, the chief officer of police. A computerised form stating that the author is so acting has been held by the Divisional Court to be sufficient for this purpose (*Arnold* v *DPP* [1999] RTR 99).

The information must be provided within a reasonable time which may, in the prevailing circumstances, mean immediately (see *Lowe* v *Lester* [1987] RTR 30).

Where the requirement is made in writing and served by post, it shall have effect as a requirement to provide the information within 28 days beginning with the day on which it is served (s. 172(7)(a) (see the defence below)).

If a person falsely claims to have been the driver, an offence of perverting the course of justice may be appropriate (**see Crime, chapter 15**).

Where the defendant is a company or corporate body, any director, secretary, manager or other officer of the company may also be prosecuted for the offence if it can be shown that the offence was committed with his/her 'consent or connivance' (s. 172(5)).

Compare the requirements of this section with those under the Vehicle Excise and Registration Act 1994 (**see chapter 12**).

The power under s. 172, which is extremely useful in a number of aspects of police work, came under attack in the run up to the Human Rights Act 1998. It was argued by some that the requirements of s. 172 infringed a defendant's right against self-incrimination under Article 6 of the European Convention on Human Rights (as to which, **see General Police Duties, chapter 2**). This argument was successfully used to get evidence obtained under s. 172 excluded (under s. 78 of the Police and Criminal Evidence Act 1984, **see Evidence and Procedure, chapter 13**) and for a while the future of the power looked uncertain. A lot of the fuss about s. 172 was started by an early decision in the Scottish courts but once that case finally reached the House of Lords, the issue was largely defused. In that case, *Brown* v *Stott* (*Procurator Fiscal, Dunfermline*) [2001] 2 WLR 817, the House of Lords held that the crucial issues were whether s. 172:

- represented a disproportionate response to the high incidence of death and injury on the roads by reason of the misuse of cars; and/or
- undermined the right to a fair trial when the driver's admission was relied on at trial.

Their Lordships held that the European Convention had to be read as balancing community rights with individual rights. The answer to both issues above was 'no' because (among other things):

- The answer required of a keeper by s. 172 could not of itself incriminate the suspect since it was not an offence merely to drive a car.
- All those who owned or drove cars had subjected themselves to a regulatory regime of which s. 172 was a part.

Athough this was a Scottish case and some of the arguments turned on the need for corroboration, the Divisional Court for England and Wales has confirmed that an admission to being the driver of a particular vehicle given in response to a s. 172 requirement does not breach the defendant's privilege against self-incrimination under Article 6 of the European Convention (*DPP* v *Wilson* [2002] RTR 6). The court went on to say that, where a defendant disputed the reliability of any such admission, a judge (or magistrate) ought to exercise their general discretion to exclude the written evidence and require the prosecution to adduce oral evidence which could then be tested by cross-examination. The court also held that there was no difference, so far as the effects of the Human Rights Act 1998 was concerned, between s. 172(2)(a) and (b).

2.9.1 Defence

Section 172(4) provides that a person shall not be convicted of an offence under s. 172(3) if that person can show that he/she did not know *and* could not have ascertained with reasonable diligence who the driver of the vehicle was.

Section 172(7)(b) provides that, where the requirement is made in writing by post, the person shall not be guilty of an offence if that person can show that either he/she gave the information as soon as reasonably practicable after the end of the 28 day period *or* that it has not been reasonably practicable for him/her to give it.

In both of these cases the evidential burden of proof lies on the defendant.

2.10 Police Drivers

Police drivers will be judged against the same standard of care as other drivers (*Wood* v *Richards* [1977] RTR 201) and there is no special exemption for them or any other emergency crews (*R* v *O'Toole* (1971) 55 Cr App R 206). In relation to the general requirements for driving with due consideration for other road users, **see para. 2.4**. However, police drivers are granted some exemptions from specific traffic regulation and there are several cases—both civil and criminal—that help in identifying the relevant features that will be applied in determining the driver's liability.

It is also worth bearing in mind that, while the particular circumstances under which the police driver found himself/herself driving may not provide a specific defence, they may nevertheless provide significant mitigation and, where appropriate, special reasons for not disqualifying the driver (see, for example, *Agnew* v *DPP* [1991] RTR 144).

The Court of Appeal's recent review of the issues of civil liability arising from police drivers (*Keyse* v *Commissioner of the Metropolitan Police* [2001] EWCA Civ 715) held that:

- Speed alone was not decisive of negligence by a police driver.
- Emergency service vehicles on duty were expressly exempted from the statutory rules for speed limits, keep left signs and traffic lights (s. 87 of the Road Traffic Regulation Act 1984 and regs 15(2) and 33(2) of the Traffic Signs Regulations and General Directions 1994 (SI 1994 No. 1519) (note the relevant regulation is now reg. 36(1)(b) of the Traffic Signs Regulations 2002 (SI 2002/3113)).
- Police and other emergency service drivers were entitled to expect other road users to take note of the signs of their approach (e.g. sirens and flashing lights) and, where appropriate, react accordingly.

Consequently, the Court held that a police driver who had accelerated through a green light at a road junction with the vehicle's visual and audible warning equipment in operation, was not liable for the injuries to a pedestrian who stepped off the pavement into the path of the vehicle.

Regulation 33 (now reg. 36(1)(b) of the Traffic Signs Regulations 2002 (SI 2002/3113)) referred to above makes allowances for emergency services drivers to pass through red traffic lights under certain circumstances. The regulations set out the conditions under which a driver may proceed through a red light and failure to meet those conditions will render the driver liable to a charge under s. 2 or 3 of the 1988 Act. Consequently, the driver of a police surveillance vehicle following suspects to the scene of an intended armed robbery could not rely on the fact that he was required by the seriousness of the circumstances to go through a red light when he did so in a way which was not covered by the regulations (*DPP* v *Harris* [1995] RTR 100). See also *R* v *Collins* [1997] RTR 439 at **para. 2.2.2**.

In *Harris*, the Divisional Court held that the wording of the regulation meant that there was no scope for the defence of necessity where police drivers went through red traffic lights. It was held that the exemption for drivers of police vehicles in such circumstances was restricted to the extent that another driver should not be obliged to change speed or course to avoid a collision. Therefore there was no scope for the common law defence (as to which, **see para. 2.11**) The court held that all the driver in *Harris* need have done to avoid the accident was to have stopped for a couple of seconds or to have edged forward slowly.

This view was followed in the civil case of *Griffin* v *Merseyside Regional Ambulance* [1998] PIQR P 34, where the Court of Appeal held that the duty of care required of a driver of any emergency vehicle is not to proceed 'in a manner likely to endanger any person' or cause them to change their speed or course of direction (per reg. 33(1)(b) (now reg. 36(1)(b) of the Traffic Signs Regulations 2002 (SI 2002/3113))). It is the *manner* of the police officer's driving that must not cause the other driver to be endangered or to change speed or direction. The whole purpose of the audible and visual warning systems on emergency vehicles ('blues and twos') is to alert other drivers and, if necessary, to get them to change speed and/or direction. Similarly, rule 76 of the Highway Code tells drivers to look and listen for emergency vehicles and to make room for them, pulling over and stopping where necessary provided that does not endanger other road users. (For the admissibility of the Highway Code's provisions, **see para. 2.8**.)

For the provisions relating to emergency service vehicles at pedestrian crossings, **see chapter 8**.

In relation to drink driving offences by police officers, the Home Office Guidance to Chief Officers has been amended to include clear guidelines on treatment and punishment (**see appendix 12**).

It is perhaps worth noting that, if a person drives in a way which creates a need for police officers to pursue him/her and an officer is subsequently injured in that pursuit, that person may owe a duty of care to the police officer and may therefore be sued for damages by the injured officer (*Langley* v *Dray* [1998] PIQR P8/314). This decision suggests that there are times when the police can be placed under an obligation by a driver to pursue him/her.

2.11 Defences

Why should police officers be concerned with defences? There are two main reasons. The first reason is because police officers have a duty to gather and present any relevant evidence whether it tends to prove or disprove a person's guilt. The second reason is because it is advisable to address any relevant defences—usually at interview stage—in order to prevent spurious defences being raised later (**see Crime, chapter 4**). (A third reason would be because police officers can often become defendants themselves in driving cases!)

The very narrow defence(s) of *duress* and *necessity* have given practitioners and academics a great deal of food for thought. Basically these common law defences apply where the defendant has been compelled to commit an offence, either by a direct threat from another person or by the prevailing circumstances. Whatever the particular arguments for and against the defences, it is clear that they will generally apply to cases of dangerous, careless and inconsiderate driving (they also apply to the offences of driving while disqualified (*R* v *Martin* [1989] RTR 63 and confirmed in *R* v *Backshall* [1998] 1 WLR 1506). Practically speaking, the defences will apply only where the driver was forced to commit the offence in order to avoid death or serious injury (*R* v *Conway* [1989] RTR 35). It is clear from *DPP* v *Harris* (**see para. 2.10**), however, that the defence is not available to police officers driving through red traffic lights in an emergency situation.

In *DPP* v *Bell* [1992] RTR 335, the defence of duress of circumstances was upheld where the defendant had needed to drive a vehicle in order to escape serious physical harm, even though he had drunk a considerable amount of alcohol. However, where the defendant had driven a far greater distance than was necessary to escape the relevant danger, the Divisional Court were not prepared to accept the defence in answer to a charge of driving while over the prescribed limit (*DPP* v *Tomkinson* [2001] RTR 38).

In *DPP* v *Hicks* [2002] EWHC 1638—a case arising out of driving while over the prescribed limit—the Administrative Court set out the following guidelines to be applied before the defence of necessity will generally be available to drivers:

- the defence is only available if the driving was undertaken to avoid consequences that could not otherwise have been avoided;
- those consequences must have been both inevitable and involved the risk of serious harm to the driver or someone else for whom he/she was responsible;
- the driver must do no more than is reasonably necessary to avoid the harm;
- the danger of so driving must not be disproportionate to the harm threatened.

The defence of 'self-defence' (**see Crime, chapter 4**) also appears to be available, where appropriate, to offences involving driving standards. This was made clear by the Court of Appeal in *R* v *Symonds* [1998] Crim LR 280 where the defendant was approached by a drunken pedestrian who tried to drag him from his car. It was held that, to allow the defence of 'self-defence' in relation to an *assault* caused by driving a vehicle (in this case, by driving off while the pedestrian had his arm through the driver's window), but not to allow it in relation to an offence such as dangerous driving arising from the same incident, was an anomaly. Therefore the defence of 'self-defence' should, where the circumstances require it, be open to someone who is charged with an offence involving the standard of their driving.

The defence afforded by the Criminal Law Act 1967, s. 3(1) (**see Crime, chapter 4**)—which allows the use of reasonable force in arresting offenders and preventing crime—may apply to cases of dangerous, careless or inconsiderate driving (*R* v *Renouf* [1986] 2 All ER 449). See also the 'defence' of automatism in **chapter 1**.

2.11.1 Mechanical Defect

If an unforeseen mechanical defect suddenly deprives the driver of control of the vehicle, he/she may have a defence similar to 'automatism' (**see chapter 1**). The reasoning for this seems to be that there is little distinction between a totally unexpected aberration in the bodily functions of a driver and a similar eventuality affecting the workings of their vehicle. In each case the driver has been deprived suddenly and unexpectedly of proper control of the vehicle and ought to be afforded the same 'defence' (*R* v *Spurge* [1961] 2 QB 205). As with automatism, if the possibility of such a defect ought to have been foreseeable by reasonable prudence, the defence will not be available.

The possibility of raising 'mechanical defect' as a defence has given rise to a number of calls from the higher courts for a procedure whereby vehicles involved in accidents cannot be scrapped without the permission of the police (see, e.g., *R* v *Thind* (1999) 19 WRTLB 12).

3 Notices of intended prosecution

3.1 Introduction

Under s. 1 of the Road Traffic Offenders Act 1988, before certain offences can be prosecuted:

- the defendant must have been warned of the possibility of that prosecution at the time of the offence (s. 1(1)(a)); or
- the defendant must have been served with a summons (or charged) within 14 days of the offence (s. 1(1)(b)); or
- a notice setting out the possibility of that prosecution must have been sent to the driver or registered keeper of the vehicle within 14 days of the offence (s. 1(1)(c)).

The notice or warning must be given by the 'prosecutor' which will ordinarily be the police. If the person giving it is not empowered to make a decision whether or not to prosecute (such as a vehicle examiner employed by the vehicle inspectorate), the warning or notice will not be deemed to have been served (*Swan* v *Vehicle Inspectorate* [1997] RTR 187).

The notice referred to is a Notice of Intended Prosecution (NIP). If a verbal warning is given at the time it must be shown that the defendant understood it (*Gibson* v *Dalton* [1980] RTR 410). Proof that they understood it will lie on the prosecution and, for that reason, it is common practice to send an NIP whether a verbal warning was given or not.

Where an offence is committed partly within the jurisdiction of one court and partly within that of another, either court may try the case. Consider the following scenario. A person whose home address is in county A commits a traffic offence in county B. Later, at their home address, that person is required to provide details of the driver at the time of the original traffic offence. The person fails to provide the details as required. The fact that the original offence was in county B does not stop the court in county A from hearing the case as the person's subsequent offence—failure to provide the details required—occurred within the jurisdiction of county A (*Kennet DC* v *Young* [1999] RTR 235).

3.2 Relevant Offences

The offences which require an NIP are listed in sch. 1 to the Road Traffic Offenders Act 1988 and include:

- Dangerous, careless or inconsiderate driving (**see chapter 2**).
- Dangerous, careless or inconsiderate cycling (**see chapter 2**).
- Failing to comply with traffic signs and directions (**see chapter 10**).
- Leaving a vehicle in a dangerous position (**see chapter 8**).
- Speeding offences under ss. 16 and 17 of the Road Traffic Regulation Act 1984 (**see chapter 7**).

The list of offences in sch. 1 is exhaustive; other offences, even if similar in nature to those in the list, will not be covered by the requirements of s. 1(1) (*Sulston* v *Hammond* [1970] 1 WLR 1164).

3.3 Exceptions

Section 2(1) of the Road Traffic Offenders Act 1988 states that the requirement to serve an NIP does not apply in relation to an offence if:

- at the time or
- immediately afterwards and
- owing to the presence of the vehicle concerned
- on a road
- an accident occurred.

Each of these features must be present to remove the need for an NIP to be served or a warning given.

'Accident' is defined at **para. 1.2.7** and is broader than the expression used under s. 170 of the Road Traffic Act 1988 (**see chapter 1**). However, although such 'reportable' accidents now extend to public places as well as roads, the exemption under s. 2(1) above is limited to an accident that occurs on *a road* at the time or immediately after the offence.

Whether the accident occurred 'at the time' is a matter of fact and degree (*R* v *Okike* [1978] RTR 489).

If the driver is unaware that the accident has taken place because it is so minor, there *will* be a need to serve an NIP (*Bentley* v *Dickinson* [1983] RTR 356). The thinking behind this ruling is that the warning or NIP will allow the driver to gather evidence to answer any charge arising from the accident.

Where the accident is so severe that the driver has no recollection of it, there is no need to serve an NIP and the ruling in *Bentley* (above) will not apply (*DPP* v *Pidhajeckyj* [1991] RTR 136).

There must be some causal connection between the presence of the vehicle concerned and the accident (*Quelch* v *Phipps* [1955] 2 All ER 302).

The vehicle must have been on a 'road' (**see para. 1.2.7**).

Section 2 goes on to say:

> (2) The requirement of section 1(1) of this Act does not apply in relation to an offence in respect of which—
> (a) a fixed penalty notice (within the meaning of Part III of this Act) has been given or fixed under any provision of that Part, or
> (b) a notice has been given under section 54(4) of this Act.
>
> (3) Failure to comply with the requirement of section 1(1) of this Act is not a bar to the conviction of the accused in a case where the court is satisfied—
> (a) that neither the name and address of the accused nor the name and address of the registered keeper, if any, could with reasonable diligence have been ascertained in time for a summons or, as the case may be, a complaint to be served or for a notice to be served or sent in compliance with the requirement, or
> (b) that the accused by his own conduct contributed to the failure.

KEYNOTE

For fixed penalty procedures generally, see chapter 14.

Failure to observe the requirements of s. 1(1) will not bar an alternative conviction which is allowed under s. 24, that is, where the original offence was not one requiring an NIP but where the alternative office *would* ordinarily require such a notice (s. 2(4)).

If the defendant contributes to the failure to serve the NIP, then that failure will not be a bar to conviction.

3.4 Presumption

Section 1 of the Road Traffic Offenders Act 1988 states:

(3) The requirement of subsection (1) above shall in every case be deemed to have been complied with unless and until the contrary is proved.

KEYNOTE

This places the burden of proving non-conformity with the requirement on the defence.

3.5 Proof

Section 1 of the Road Traffic Offenders Act 1988 states:

(1A) A notice required by this section to be served on any person may be served on that person—
- (a) by delivering it to him;
- (b) by addressing it to him and leaving it at his last known address; or
- (c) by sending it by registered post, recorded delivery service or first class post addressed to him at his last known address.

(2) A notice shall be deemed for the purposes of subsection (1)(c) above to have been served on a person if it was sent by registered post or recorded delivery service addressed to him at his last known address, notwithstanding that the notice was returned as undelivered or was for any other reason not received by him.

(3) The requirement of subsection (1) above shall in every case be deemed to have been complied with unless and until the contrary is proved.

KEYNOTE

As discussed above, the warning may be oral (under s. 1(1)(a)) but there are problems of proof where such warnings are given. If the defendant was preoccupied at the time or can show that he/she did not fully understand what was being said, the provisions of s. 1(1) will not have been complied with.

Serving an NIP personally on the spouse or partner of the defendant would appear to be enough (*Hosier* v *Goodall* [1962] 1 All ER 30).

If the defendant is not at his/her home address, for instance because he/she is in hospital or on holiday, service to his/her last known address will suffice even if the police are aware of that fact (*Phipps* v *McCormick* [1972] Crim LR 540).

If neither the defendant nor the registered keeper have any fixed abode, reasonable efforts must be made to serve the notice personally. If such efforts fail, s. 2(3) would apply and the need for service would be removed.

As the purpose of the warning or notice is to alert the defendant to the likelihood of prosecution, it is not necessary to specify exactly which offence is being considered; it is enough that the defendant is made aware of the *nature* of the offence (*Pope* v *Clarke* [1953] 2 All ER 704).

4 Accidents and collisions

4.1 Introduction

Although there are many types of event involving vehicles which can be described as an 'accident'—including those brought about deliberately (**see chapter 1**)—there are specific circumstances where that 'accident' imposes duties on certain people involved. This chapter is primarily concerned with these 'reportable' accidents.

4.2 Reportable Accidents

Section 170 of the Road Traffic Act 1988 states:

(1) This section applies in a case where, owing to the presence of a mechanically propelled vehicle on a road or other public place, an accident occurs by which—

(a) personal injury is caused to a person other than the driver of that mechanically propelled vehicle, or

(b) damage is caused—

(i) to a vehicle other than that mechanically propelled vehicle or a trailer drawn by that mechanically propelled vehicle, or

(ii) to an animal other than an animal in or on that mechanically propelled vehicle or a trailer drawn by that mechanically propelled vehicle, or

(iii) to any other property constructed on, fixed to, growing in or otherwise forming part of the land on which the road or place in question is situated or land adjacent to such land.

(2) The driver of the mechanically propelled vehicle must stop and, if required to do so by any person having reasonable grounds for so requiring, give his name and address and also the name and address of the owner and the identification marks of the vehicle.

(3) If for any reason the driver of the mechanically propelled vehicle does not give his name and address under subsection (2) above, he must report the accident.

(4) ...

(5) If, in a case where this section applies by virtue of subsection (1)(a) above, the driver of a motor vehicle does not at the time of the accident produce such a certificate of insurance or security, or other evidence, as is mentioned in section 165(2)(a) of this Act—

(a) to a constable, or

(b) to some person who, having reasonable grounds for so doing, has required him to produce it,

the driver must report the accident and produce such a certificate or other evidence.

This subsection does not apply to the driver of an invalid carriage.

(6) To comply with a duty under this section to report an accident or to produce such a certificate of insurance or security, or other evidence, as is mentioned in section 165(2)(a) of this Act, the driver—

(a) must do so at a police station or to a constable, and

(b) must do so as soon as is reasonably practicable and, in any case, within twenty-four hours of the occurrence of the accident.

(7) ...

(8) In this section 'animal' means horse, cattle, ass, mule, sheep, pig, goat or dog.

KEYNOTE

Accidents as defined under s. 170(1)(a) and (b) above are generally referred to as 'reportable' accidents. These should be distinguished from the wider definition of 'accident' used elsewhere in the Act (**see para. 1.2.7**).

Following the decision of the House of Lords in *Cutter* v *Eagle Star Insurance Co. Ltd* [1998] 4 All ER 417 (as to which, **see chapter 1**), the requirements under s. 170 were extended to cover accidents occuring in 'other public places' as well as roads (Motor Vehicles (Compulsory Insurance) Regulations 2000 (SI 2000 No. 726)).

Note that the duty upon the driver under s. 170(2) is to stop *and* give his/her name and address (and name and address of the owner) (*DPP* v *Bennett* [1993] RTR 175).

However, it has been held that, as the reason for this requirement to 'exchange details' is to allow future communications between interested parties, the address of the driver's solicitor would be enough to discharge the duty at s. 170(2) (*Dpp* v *McCarthy* [1999] RTR 323).

No mention is made of who may have been to blame for the accident. A driver will attract the duties imposed by s. 170 even if he/she is not at fault. In order to attract those duties, however, the driver must know of the accident (see *Harding* v *Price* [1948] 1 All ER 283).

Even though there may be a break in the actual driving of the vehicle, the driver may still be under the obligations imposed by s. 170 if an accident occurs while he/she is away from the vehicle. Therefore, where a driver left his vehicle on a road with its hazard warning lights on while he ran to a post box and the vehicle coasted downhill into a wall, the driver still had a duty to report the accident under s. 170 (*Cawthorn* v *DPP* [2000] RTR 45). In *Cawthorn*, even the fact that a passenger may have been directly responsible for the accident (by letting off the handbrake), it was held that the circumstances met the requirements of s. 170 and that the driver attracted the relevant statutory duties. The Divisional Court held that the offence under s. 170(4) is not a 'driving offence' as such and does not require that the person be 'driving' at the time.

In proving the offence you would not have to show that the driver was so aware; once the damage or injury is proved it is for the driver to show (on the balance of probabilities) that he/she was unaware of its occurrence (*Selby* v *Chief Constable of Avon and Somerset* [1988] RTR 216).

Damage caused under the circumstances set out at s. 170(1)(b) applies to any vehicle, not just one which is mechanically propelled (such as bicycles and trailers).

If the only damage caused is to the vehicle concerned or an animal in it then the accident will not fall within the criteria of being 'reportable'. Similarly if the only person injured is the driver of the vehicle concerned, the accident will not be reportable; it will, however, be reportable if a passenger in that vehicle is injured.

Where a bus driver braked hard, causing injury to one of his passengers, the Divisional Court held that he had a duty to stop at the place and the time that the passenger was caused to be injured. It was not open to the driver to claim that the bus itself was the scene of the accident. The fact that the driver failed to stop immediately meant that he was guilty of an offence under s. 170(4) (**see para. 4.3**) (*Hallinan* v *DPP* [1998] Crim LR 754).

The list of animals under s. 170(8) does not include one of the most common 'victims' of road traffic accidents—cats.

The accident must have been brought about owing to the vehicle's presence on a 'road' or other public place (**see chapter 1**); therefore s. 170 does not apply to accidents which occur on private premises.

In order to determine whether the motor vehicle was being driven without insurance (**see chapter 6**), s. 171 of the Road Traffic Act 1988 places a requirement on the *owner* of the vehicle to give such information as required the police. Failure to comply with such a requirement is a summary offence.

Where it appears to a constable that, by reason of *any* accident having occurred owing to the presence of the vehicle on a road, it is requisite that a test should be carried out forthwith, he/she may require it to be so carried out and, if the officer does not carry out the test himself/herself, he/she can require that the vehicle is not taken away until the test has been carried out (s. 67(7) of the Road Traffic Act 1988). For testing of vehicles generally, **see chapter 9**.

4.2.1 Stop

Stop under s. 170(2) means to stop and remain at the scene for such a time as would allow anyone having a right or reason for doing so to ask for information from the driver (*Lee* v *Knapp* [1967] 2 QB 442); it does not require the driver to make enquiries of his/her own to try and find such a person (see *Mutton* v *Bates* [1984] RTR 256).

This duty also appears to require the driver to stay with his/her vehicle for a reasonable time (*Ward* v *Rawson* [1978] RTR 498).

4.2.2 Injury

Injury under s. 170 has been held to include shock and, given the recent developments in the area of assaults (e.g. *R* v *Ireland* [1997] 3 WLR 534; *R* v *Chan-Fook* [1994] 1 WLR 689 (**see Crime, chapter 8**)), it would appear that psychological harm may well amount to an 'injury' for these purposes.

4.2.3 Duty to Report

If the driver does not give his/her name and address, he/she must report the accident to a police officer or at a police station. This must be done as soon as reasonably practicable (which will be a question of fact for the court to decide in each circumstance) and in any case within 24 hours of the occurrence of the incident. This latter requirement does not give the driver up to 24 hours to report the accident; that report must be made as soon as is reasonably practicable. The report to the police within 24 hours appears to apply even if it is not practicable—or even possible?—to do so within that time limit (see *Bulman* v *Lakin* [1981] RTR 1).

That duty is probably not discharged if the driver waits until the police call at his/her house and then tells them (see *Dawson* v *Winter* (1932) 49 TLR 128 *obiter*); neither is it discharged if the driver tells a friend who happens to be a police officer (*Mutton* v *Bates* [1984] RTR 256).

Any report must be made in person; telephoning (and presumably sending a fax or e-mail) a police station is not enough (*Wisdom* v *McDonald* [1983] RTR 186).

4.3 The Offences

OFFENCE: **Failing to Stop or Report an Accident—*Road Traffic Act 1988, s. 170(4)***

- Triable summarily • Six months' imprisonment and/or fine
- Discretionary disqualification

(If personal injury caused to person other than driver, arrestable offence; otherwise no specific power of arrest)

The Road Traffic Act 1988, s. 170 states:

> (4) A person who fails to comply with subsections (2) or (3) above is guilty of an offence.

OFFENCE: **Failing to Produce Proof of Insurance after Injury Accident—*Road Traffic Act 1988, s. 170(7)***

- Triable summarily • Fine

(No specific power of arrest)

The Road Traffic Act 1988, s. 170 states:

> (7) A person who fails to comply with a duty under subsection (5)... is guilty of an offence, but he shall not be convicted by reason only of a failure to produce a certificate or other evidence if, within seven days after the occurrence of the accident, the certificate or other evidence is produced at a police station that was specified by him at the time when the accident was reported.

KEYNOTE

This is not a 'driving' offence and s. 170 is not limited to situations where the accident is caused by 'driving' the vehicle concerned (see *Cawthorn* above). Whether a defendant was the driver of a relevant vehicle at the relevant time is a question of fact for the court to determine. In doing so, the court may take into account facts such as those in *Cawthorn*, e.g. that the defendant had been driving the vehicle before the accident, that he had parked the vehicle—returning to it immediately after the accident—and that he had left his hazard warning lights on indicating that his journey was not finished.

If a driver fails to stop *and* fails to report an accident he/she commits *two* offences (*Roper* v *Sullivan* [1978] RTR 181).

From 1 October 2001 the offence of failing to stop after a personal injury accident (so called 'hit and run' incidents) is an arrestable offence (as to which, **see General Police Duties, chapter 2**)—see s. 71 of the Criminal Justice and Police Act 2001. This not only increases the powers of arrest available, but also powers of entry where an offence is suspected. These powers of entry may be particularly useful in drink driving cases where the law is unclear (**see para. 5.4.4**).

The Court of Appeal refused to accept that a driver who failed to stop and report an accident for fear of being breathalysed did not, without more, commit an offence of perverting the course of justice (*R* v *Clark* [2003] EWCA Crim 991). While it is true that driving home and remaining there allowed the passage of time to dissipate the level of alcohol in the driver's body and thereby avoid conviction for a drink driving offence (as to which, **see chapter 5**), the Court was not convinced that this was enough to justify being charged with such a serious offence. The *failure* to report the accident could not amount to an offence of perverting the course of justice as that offence required a positive *act* by the defendant. For the offence of perverting the course of justice and similar offences, **see Crime, chapter 15**.

OFFENCE: **Refusing to Give, or Giving False Details after Allegation of Dangerous or Careless Driving or Cycling—*Road Traffic Act 1988, s. 168***

• Triable summarily • Fine

(No specific power of arrest)

The Road Traffic Act 1988, s. 168 states:

Any of the following persons—

(a) the driver of a mechanically propelled vehicle who is alleged to have committed an offence under section 2 or 3 of this Act, or

(b) the rider of a cycle who is alleged to have committed an offence under section 28 or 29 of this Act,

who refuses, on being so required by any person having reasonable ground for so requiring, to give his name or address, or gives a false name or address, is guilty of an offence.

KEYNOTE

This requirement arises where the driver (or rider) is alleged to have committed an offence of dangerous or careless driving or cycling (**see chapter 2**).

There is no requirement for an accident of any sort to have occurred.

The section uses the word 'refuses' but makes no mention of a *failure* to provide the required details. Although s. 11(2) of the 1988 Act makes provision for a 'failure' to include a 'refusal' (in relation to drink driving offences), the Act says nothing about a *vice versa* situation. Given that omission, together with the fact that the courts have held (albeit in an employment law case) that 'failure' is not synonymous with 'refusal' (see *Lowson* v *Percy Main & District Social Club and Institute Ltd* [1979] ICR 568), it would seem that a mere *failure* to provide the details required under s. 168 will not amount to the offence above.

5 Drink driving

5.1 Introduction

The law regulating drinking and driving is contained in the Road Traffic Act 1988 and the Road Traffic Offenders Act 1988. It has been substantially developed by common law and entire books could be devoted to the cases which the subject has generated since its introduction over 30 years ago.

One of the good things about the subject, both from a practical point of view, and that of the student, is that virtually every conceivable circumstance, excuse or mitigating situation has been put before the courts. The result is that, when approaching a set of real or hypothetical circumstances under the drink/drive laws, the chances are that it has already been tried—both literally and judicially!

Although what follows in this chapter is a discussion of some of the relevant law and powers of enforcement, there is an increasing emphasis on judicial action to prevent re-offending. The Road Traffic Offenders Act 1988 contains a series of measures designed to allow for driver retraining and to offer reduced periods of disqualification for drivers who undergo the prescribed courses. The initial period of disqualification must be at least 12 months; any reduction cannot be less than three months or more than a quarter of the whole period (see s. 34A). Drivers must be at least 17 years old to take part in the scheme and he/she must generally pay for the course. Full details of the courses are set out in the various regulations made under s. 34C.

5.2 Unfit Through Drink or Drugs

OFFENCE: **Driving or Attempting to Drive a Mechanically Propelled Vehicle when Unfit Through Drink or Drugs—*Road Traffic Act 1988, s. 4(1)***

- Triable summarily • Six months' imprisonment and/or a fine
- Obligatory disqualification

(Statutory power of arrest)

The Road Traffic Act 1988, s. 4 states:

> (1) A person who, when driving or attempting to drive a mechanically propelled vehicle on a road or other public place, is unfit to drive through drink or drugs is guilty of an offence.

OFFENCE: **Being in Charge of a Mechanically Propelled Vehicle when Unfit Through Drink or Drugs—*Road Traffic Act 1988, s. 4(2)***

- Triable summarily • Three months' imprisonment and/or a fine
- Discretionary disqualification

(Statutory power of arrest)

The Road Traffic Act 1988, s. 4 states:

> (2) Without prejudice to subsection (1) above, a person who, when in charge of a mechanically propelled vehicle which is on a road or other public place, is unfit to drive through drink or drugs is guilty of an offence.

KEYNOTE

Section 4 creates three separate offences. For the definitions of 'driving', 'attempting to drive' and 'in charge', see chapter 1.

For the corresponding legislation applying to railways and tram systems, **see General Police Duties, chapter 13**.

Under s. 4:

> (3) ...a person shall be deemed not to have been in charge of a mechanically propelled vehicle if he proves that at the material time the circumstances were such that there was no likelihood of his driving it so long as he remained unfit to drive through drink or drugs.

KEYNOTE

The onus is on the defendant to prove this fact. However, in determining whether the defendant was likely to drive the vehicle while unfit a court may disregard any injury to the defendant or damage to the vehicle (s. 4(4)).

It has been held that such an onus of proof does not breach a defendant's right to a presumption of innocence (per Article 6(2) of the European Convention on Human Rights (presumption of innocence, see General Police Duties, chapter 2)—*Sheldrake* v *DPP* [2003] 2 All ER 497. Although the wording of the above defence means the issues are slightly different here from those under s. 5(1)(b), the requirements and standards of proof involved are the same (see para. 5.3.1).

For the definitions of 'mechanically propelled vehicle', 'road' and 'public place', see chapter 1.

Evidence

Evidence of impairment must be produced by the prosecution. That evidence may be given by any 'lay' witness and does not require expert testimony (*R* v *Lanfear* [1968] 2 QB 77). Such a witness cannot, however, give evidence as to the defendant's ability to drive.

It is not necessary to show what quantity of alcohol or drug the defendant had in his/her system for this offence. Therefore there is no need for any form of breath test. However, the police powers under s. 4 do not 'fall away' if you do embark on the procedure under s. 6 (e.g. requiring a breath sample: **see para. 5.4**). This was made clear in the decision of the Administrative Court in *DPP* v *Robertson* [2002] RTR 383. In that case, the officers had breathalysed the defendant who produced a negative result. However, on talking to the officers further, the defendant slurred his speech when giving the name of his solicitor. The officers then arrested the defendant under s. 4 (above) and took him to a police station where he provided an evidential sample of breath which was over the prescribed limit. The magistrates held that the defendant had been unlawfully arrested and the prosecutor appealed on a number of grounds. The Administrative Court held that it was quite conceivable to have a case where, notwithstanding that a driver had given a negative screening breath test, he/she was seen moments later staggering in a way that gave rise to a suspicion of unfitness. In such a case the law (s. 4) clearly gave a constable a power to arrest if what he/she had witnessed amounted to reasonable cause to suspect that the person was impaired.

'Drugs' will include any intoxicant other than alcohol (s. 11); toluene found in some glues will amount to such a drug (*Bradford* v *Wilson* (1983) 78 Cr App R 77). (Contrast this situation with other offences of 'drunkenness'; **see General Police Duties, chapter 4**.)

5.2.1 Police Powers

A constable may arrest a person without warrant if he/she has reasonable cause to suspect that the person is, or has been committing an offence under s. 4 (s. 4(6)). For the purpose of arresting a person under the power above, a constable may enter (if need be by force) any place where that person is or where he/she, with reasonable cause, suspects them to be (s. 4(7)).

The suspicion need not be formed while the defendant is driving and can become apparent after he/she has been stopped or spoken to for some other reason (*R* v *Roff* [1976] RTR 7).

Note that, in both the power to arrest and the power to enter, the suspicion required is the lesser one of reasonable cause to *suspect*; contrast reasonable cause to *believe* (**para. 5.4.1**). In *Swales* v *Cox* [1981] QB 849 the Divisional Court held that officers would need to show that force was actually necessary before its use could be justified and therefore it would be prudent in every case to seek admittance first. The courts will be even keener to monitor this aspect since the Human Rights Act 1998 came into force.

5.3 Over Prescribed Limit

OFFENCE: **Driving or Attempting to Drive a Motor Vehicle while Over the Prescribed Limit—*Road Traffic Act 1988, s. 5(1)(a)***

- Triable summarily • Six months' imprisonment and/or a fine
- Obligatory disqualification

(Statutory power of arrest)

The Road Traffic Act 1988, s. 5 states:

(1) If a person—
  (a) drives or attempts to drive a motor vehicle on a road or other public place . . . after consuming so much alcohol that the proportion of it in his breath, blood or urine exceeds the prescribed limit he is guilty of an offence.

OFFENCE: **Being in Charge of a Motor Vehicle while Over the Prescribed Limit—*Road Traffic Act 1988, s. 5(1)(b)***

- Triable summarily • Three months' imprisonment and/or a fine
- Discretionary disqualification

(Statutory power of arrest)

The Road Traffic Act 1988, s. 5 states:

(1) If a person—
  (a) . . .
  (b) is in charge of a motor vehicle on a road or other public place,
  after consuming so much alcohol that the proportion of it in his breath, blood or urine exceeds the prescribed limit he is guilty of an offence.

KEYNOTE

These offences apply to *motor vehicles* as defined in **para. 1.2.1**; contrast the previous offences in relation to unfitness to drive.

'Consuming' is not restricted to drinking and will encompass other methods of getting alcohol into the bloodstream (*DPP* v *Johnson* [1995] RTR 9).

Where a person is charged under s. 5(1)(a), the charge should specify whether the person was 'driving' or 'attempting to drive' (*R* v *Bolton Justices, ex parte Zafer Alli Khan* [1999] Crim LR 912).

The prescribed limit (at the time of writing) is:

- 35 microgrammes of alcohol in 100 millilitres of breath
- 80 milligrammes of alcohol in 100 millilitres of blood
- 107 milligrammes of alcohol in 100 millilitres of urine.

Section 11(2) which sets out these limits also allows for the levels to be changed and is under review. At the time of writing there was a suggestion that the limit be reduced for new drivers (**see para. 11.8**).

The prosecution have to establish that the defendant's breath-alcohol content exceeded the permitted maximum; it does not have to establish a specific figure (*Gordon* v *Thorpe* [1986] RTR 358).

When dealing with the offence under s. 5(1)(b) it will be important to remember the statutory defence (**see para. 5.3.1**). The effect of this is that police officers investigating offences of being 'in charge' will have to ensure that a defendant is interviewed properly and fully as to all the circumstances, taking particular care to establish that he/she is fully sober at the time of interview and that there was a real risk of his/her driving the vehicle while still over the prescribed limit.

5.3.1 Defence

Section 5 goes on to state:

> (2) It is a defence for a person charged with an offence under subsection (1)(b) above to prove that at the time he is alleged to have committed the offence the circumstances were such that there was no likelihood of his driving the vehicle whilst the proportion of alcohol in his breath, blood or urine remained likely to exceed the prescribed limit.
>
> (3) The court may, in determining whether there was such a likelihood as is mentioned in subsection (2) above, disregard any injury to him and any damage to the vehicle.

KEYNOTE

This defence was subjected to considerable scrutiny in a recent case where a defendant was found asleep in a van and, on being breathalysed, provided a reading that showed his alcohol to breath ratio to be four times the legal limit. The defendant claimed that there had been no likelihood of his driving while over the prescribed limit but the magistrates did not accept that he had proved the point sufficiently. The defendant then argued that the application of s. 5(2) above was contrary to Article 6(2) of the European Convention on Human Rights (presumption of innocence, **see General Police Duties, chapter 2**). Although the Divisional Court failed to agree at the first hearing and were still not unanimous on the second occasion, the majority decision was that the defence could be read as follows: 'It is a defence for a person charged with an offence under s. 5(1)(b) above to *demonstrate from the evidence an arguable case that* at the time he is alleged to have committed the offence etc.' (emphasis added). Therefore, once the prosecution prove the elements of the offence beyond a reasonable doubt, the defendant must demonstrate an arguable case that there was no likelihood of his/her driving the vehicle while still over the prescribed limit. Once the defendant does this, the prosecution must then prove beyond a reasonable doubt that there *was* such a likelihood (*Sheldrake* v *DPP* [2003] 2 All ER 497).

For the defence of duress of circumstances or necessity, **see Crime, chapter 4**.

5.4 Preliminary Breath Tests

Section 6 of the Road Traffic Act 1988 provides for the administering of preliminary breath tests, sometimes referred to as 'screening' or 'roadside' breath tests (as opposed to evidential breath tests (**see para. 5.5**)). The procedures for obtaining such breath tests are summarised in **appendix 8, diagrams 1 and 2**.

If officers choose to administer a breath test under this section, their police powers under s. 4 (**see para. 5.2**) are still available under appropriate circumstances (*DPP* v *Robertson* [2002] RTR 383).

Section 6 of the Road Traffic Act 1988 states:

(1) Where a constable in uniform has reasonable cause to suspect—
   (a) that a person driving or attempting to drive or in charge of a motor vehicle on a road or other public place has alcohol in his body or has committed a traffic offence whilst the vehicle was in motion, or
   (b) that a person has been driving or attempting to drive or been in charge of a motor vehicle on a road or other public place with alcohol in his body and that that person still has alcohol in his body, or
   (c) that a person has been driving or attempting to drive or been in charge of a motor vehicle on a road or other public place and has committed a traffic offence whilst the vehicle was in motion,

   he may, subject to section 9 of this Act, require him to provide a specimen of breath for a breath test.

KEYNOTE

The breath test referred to is 'for the purpose of obtaining, by means of a device of a type approved by the Secretary of State, *an indication* whether the proportion of alcohol in a person's breath or blood is likely to exceed the prescribed limit' (s. 11(2)—emphasis added). This is an important distinction as the preliminary breath test is undertaken only to give an indication to the officer administering it of the likelihood of the offence, not for proving the relevant offence.

However, evidence of a roadside breath test is admissible on the issue of the defendant's veracity, i.e. to prove whether he/she is telling the truth or not (see *DPP* v *Brown & Teixeira* [2002] RTR 23).

Failure to comply with the manufacturer's instructions on the use of approved devices (e.g. when assembling the tube on the Lion Alcolmeter or allowing the driver to smoke immediately before taking the test) will mean that the person has *not provided* a preliminary breath test and may be asked to provide another; refusing to do so will be an offence (see below) (*DPP* v *Carey* [1969] 3 All ER 1662).

Note that a driver cannot gain an extra 20 minutes' grace by pretending that they have just had a drink (see *Grant* v *DPP* [2003] EWHC 130).

The Divisional Court decided recently that an innocent failure by a police officer to follow the manufacturer's instructions should not be deemed to render either the test or any subsequent arrest unlawful (*DPP* v *Kay* [1999] RTR 109).

The procedure specifies that a constable requiring the test must be in uniform (contrast the powers available under s. 4 above). The section does not specifically mention who may *administer* the test once the requirement has been made. At the time of writing there is no reliable authority on the point. When deciding on the issue of whether some other officer, in uniform or otherwise, can actually *administer* a breath test that has been lawfully required under s. 6(1), the following arguments may be brought:

- Parliament clearly intended to provide certain safeguards in relation to the taking of breath specimens in certain situations, under this and the previous legislation. The inclusion at s. 6(1) of the requirement for an officer to be in uniform when making the request must be seen as a deliberate restriction on the scope of that power, particularly when seen alongside the next subsection relating to accidents (**see para. 5.4.2**). The practical effect of such a safeguard would be considerably diluted if it was to be interpreted as allowing the actual *administration* of the breath test to be carried out by any other officer. If this were the case then, taken to its logical extreme, a plain clothes officer wishing to breathalyse someone sitting in a car in a public place might simply get one of their uniformed colleagues to make the request to the driver over their personal radio or by mobile phone! This cannot have been Parliament's intention.
- The courts *may* conclude that Parliament's assumption was that the officer making the request and the officer administering the test would be one and the same person, i.e. in the same way that they interpreted the former provisions of s. 5 of the Public Order Act 1986 (where the warning had to be given by the same officer who subsequently went on to arrest the defendant), necessitating the passing of the Public Order (Amendment) Act 1996 (**see General Police Duties, chapter 4**).

In the absence of an authority on this point, it is suggested that the officer administering the test under s. 6(1) should at least be in uniform when doing so (although the provisions of s. 15(2) of the Road Traffic Offenders Act 1988 means that any evidence obtained after an inadvertently flawed process may still be admissible (see **para. 5.5.2**).

Whether the constable (which includes an officer of any rank, including special constables) was in uniform at the time is a matter of fact for a court to determine in each case. What counts as uniform is unclear but, if the constable can be easily identified from his/her manner of dress as a police officer, the requirement has probably been met (*Wallwork* v *Giles* [1970] RTR 117; officer without helmet on held to be 'in uniform').

No specific form of words is needed; the officer must clearly convey a 'requirement' to take the preliminary test (*R* v *O'Boyle* [1973] RTR 445).

A traffic offence is generally any offence under the Road Traffic Act 1988 (except those relating to driving instructors), the Road Traffic Regulation Act 1984 and the Road Traffic Offenders Act 1988 (except the provisions of Part III which deal with fixed penalty procedures) (s. 6(8)). Such offences, which include offences under any regulations made under those Acts, must have been committed where the vehicle was moving.

The courts have accepted that the police are empowered to *stop* vehicles at random (under the Road Traffic Act 1988, s. 163), whether to establish the identity of the driver, to enquire whether the driver has been drinking or to train newly appointed officers in traffic procedures (see *Chief Constable of Gwent* v *Dash* [1986] RTR 41). That does not empower officers to conduct *random breath tests* but, having randomly stopped a vehicle, they may carry out preliminary breath tests provided they follow the procedure set out above. In a Scottish case it was held that the only limit on the power under s. 163 is that it should not be used capriciously or oppressively (*Stewart* v *Crowe* 1999 SLT 899).

5.4.1 Reasonable Cause to Suspect

As mentioned above (**para. 5.2.1**), there is a distinction between reasonable cause to 'suspect' and reasonable cause to 'believe' (see s. 6(2), **para. 5.4.2**).

Whether the constable had reasonable cause to suspect or believe will be a question of fact, to be determined in the light of all the available facts. The distinction between the two however, has been held to be a significant one, intended by Parliament (*Baker* v *Oxford* [1980] RTR 315). In that case it was held that the deliberate use of the word 'believe' in the Act imposed a requirement for a greater degree of certainty in the mind of the officers concerned.

Reasonable cause to suspect may arise from the behaviour of a driver but single errors of judgment or carelessness will not necessarily justify a suspicion of alcohol (*Williams* v *Jones* [1972] RTR 4).

The importance of whether such grounds were apparent or not lies, not only in triggering the relevant powers; they also affect the nature of any conversation that might take place between the officer and a suspected driver. A good example of this can be seen in *Ortega* v *DPP* [2001] EWHC Admin 143). In that case the officers were held not to have had reasonable grounds to suspect that any offence had been committed at the time they questioned the defendant about his ownership and driving of a particular car. The absence of such reasonable grounds to suspect an offence meant that the doorstep conversation which was not contemporaneously recorded and some of which took place without caution, was nevertheless admissible.

In addition to first-hand observation of a driver's behaviour, reasonable cause to suspect may arise from the observations of another officer (*Erskine* v *Hollin* [1971] RTR 199); it may even arise from information provided by a member of the public (see *DPP* v *Wilson* [1991] RTR 284). An officer receiving a radio message that a driver had been seen driving erratically may thereby acquire reasonable cause to suspect that the driver has committed an offence in relation to alcohol consumption (*R* v *Evans* [1974] RTR 232).

Reasonable cause to *believe* requires something more than suspicion (*R* v *Forsyth* [1997] Crim LR 589). Additionally, it is not enough to show that there are such grounds for the belief; the officer must actually hold that belief as well (*R* v *Banks* [1916] 2 KB 621).

The suspicion required of a constable under s. 6(1)(b) relates both to the fact that the person drove etc. with alcohol in their body *and* that they still have alcohol in their body at the time of the request.

Whether a test was carried out 'at or near' the place where the requirement was made is a question of fact for the court to determine in each case and, given previous decisions, it is unlikely that the magistrates' decision will be overruled by a higher court (**see Evidence and Procedure, chapter 2**). Note that there is no general power of entry in order to administer preliminary breath tests (**see para. 5.4.4**).

For further discussion of the general statutory meaning of 'reasonable cause to suspect', **see General Police Duties, chapter 2**.

5.4.2 Procedure Following an Accident

Section 6(2) states:

> (2) If an accident occurs owing to the presence of a motor vehicle on a road or other public place, a constable may, subject to section 9 of this Act, require any person who he has reasonable cause to believe was driving or attempting to drive or in charge of the vehicle at the time of the accident to provide a specimen of breath for a breath test.

KEYNOTE

The power under s. 6(2) is expressly subject to the provisions of s. 9 which deals with hospital patients (**see para. 5.5.4**). The specimen after an accident may be taken at or near the place where the requirement is made or, if the officer making the requirement thinks fit, at a police station specified by that officer.

The officer must show that an accident had taken place, not simply that he/she believed that to be the case (*Chief Constable of West Midlands Police* v *Billingham* [1979] RTR 446). For the definition of 'accident' in this context, **see para. 1.2.7**. The meaning is not restricted to that given to a 'reportable' accident under s. 170 (**see para. 4.2**).

The requirement here is for the officer to have 'reasonable cause to believe' that the person was driving or in charge of a relevant vehicle at the time of an accident (s. 6(2) and see *Johnson* v *Whitehouse* [1984] RTR 38). The issue of whether the officer had reasonable cause to *believe* that the person was so driving or in charge of the vehicle at the time of the accident will be a question of fact.

There is no need for the police officer making the requirement to suspect or believe that the driver has been drinking, nor that he/she has committed any offence; reasonable belief in his/her involvement in the accident is enough.

OFFENCE: **Failing to Provide Breath Specimen for Preliminary Test—*Road Traffic Act 1988, s. 6(4)***

- Triable summarily • Fine • Discretionary disqualification

(Statutory power of arrest)

The Road Traffic Act 1988, s. 6 states:

> (4) A person who, without reasonable excuse, fails to provide a specimen of breath when required to do so in pursuance of this section is guilty of an offence.

KEYNOTE

An example of 'reasonable excuse' would be where the defendant is physically unable to provide a specimen or where to do so would entail a substantial risk to his/her health (see *R* v *Lennard* [1973] 2 All ER 831).

The breath sample must be provided in such a way as to 'enable the objective of the test or analysis to be satisfactorily achieved' (s. 11(3) of the Road Traffic Act 1988). Therefore, if the person does not follow the instructions given or provides a sample in a way which does not fulfil the requirements of s. 11(3), he/she has 'failed' to provide.

Failing includes a refusal (s. 11(2)). Having been required to provide a specimen of breath, a person may also be required to wait until a device is brought to the scene; failing to wait for a reasonable time, or doing anything which demonstrates an intention not to provide the specimen can amount to a 'failure' (see *R* v *Wagner* [1970] Crim LR 535).

Simply producing enough breath to enable the device to give a positive reading does not necessarily mean the defendant has 'provided a specimen'; he/she must produce sufficient breath in the manner required to enable the device to give a reliable reading—positive or negative (*DPP* v *Heywood* [1998] RTR 1).

5.4.3 Trespassing

Many legal arguments have raged over the legitimacy of a breath test carried out, or requested where the officers concerned are in fact trespassing on the property of another. The results of these deliberations are:

- If police officers are trespassing *on the defendant's property* they are not entitled to require a breath test (*R v fox* [1986] AC 281).
- If the officers are trespassing at the time they make the requirement, any subsequent arrest made by them is unlawful (*Clowser* v *Chaplin* [1981] RTR 317).
- Although the entry and the arrest may be unlawful, this will only affect the offence under s. 6(4) and not any other offences detected at the police station (see below).
- Any requirement for a sample of breath properly made, and any subsequent, arrest remains lawful *until* the officers become trespassers. A police officer, like any other member of the public, has an implied licence to go onto certain parts of someone's property (e.g. the front doorstep) unless and until that licence is withdrawn. If a police officer goes onto such a part of the defendant's property, he/she is not trespassing until told to leave and any requirement for a breath test, or any arrest made before this happens will be lawful. (See *Pamplin* v *Fraser* [1981] RTR 494 where the defendant drove onto his own land and locked himself in his car. It was held that the officers' licence to enter the land had not been withdrawn at that stage.) Telling officers to 'fuck off' is not necessarily enough to withdraw this implied licence (*Snook* v *Mannion* [1982] RTR 321).
- A person cannot seek refuge on someone else's land as a trespasser themselves; a requirement for a breath test and any subsequent arrest may be lawful if the person is a trespasser (see *Morris* v *Beardmore* [1980] RTR 321).
- If a person has been lawfully arrested on their own land, officers may remain on the land and enter premises to recapture them, even if asked to leave (*Hart* v *Chief Constable of Kent* [1983] Crim LR 117).
- Although the provisions of s.15(2) of the Road Traffic Offenders Act 1988 may render evidence from a specimen taken after an unlawful arrest admissible (**see para. 5.5.2**), any finding that the officer(s) concerned acted in a way that they know to be unlawful or unwarranted may lead to that evidence being excluded under s. 78 of the Police and Criminal Evidence Act 1984 (see *DPP* v *Godwin* [1991] RTR 303).

5.4.4 Powers of Entry

Clearly the above situations can be avoided where there is a power of entry. In addition to that provided under s. 4 (**see para. 5.2.1**), there is a specific power of entry provided by

s. 6(6) which states:

> (6) A constable may, for the purpose of requiring a person to provide a specimen of breath under subsection (2) above in a case where he has reasonable cause to suspect that the accident involved injury to another person or of arresting him in such a case under subsection (5) above, enter (if need be by force) any place where that person is or where the constable, with reasonable cause, suspects him to be.

KEYNOTE

The wording of this subsection appears to limit the purposes of entry to:

- carrying out the test
- arresting the person following such a test
- arresting a person after they have failed to provide a specimen for such a test.

In order for this power to apply you must show that:

- **there was an accident** (mere suspicion or belief by the officer, however strong, is not enough);
- **the officer had reasonable cause to *believe* that the person had been driving, attempting to drive or in charge of the vehicle concerned** (suspicion is not enough);
- **the officer had reasonable cause to *suspect* that the accident involved injury to another person** (for this part suspicion is enough provided there is reasonable cause for it);
- ***the officer had reasonable cause to suspect* the person to be in the place entered** (again suspicion is enough).

However, because failing to stop after a personal injury accident is now an arrestable offence **(see para. 4.3)** the practical significance of these limitations may have been reduced.

5.4.5 Power of Arrest

Under s. 6(5)(a) a police officer may (not *must*) arrest a person if, as a result of a preliminary breath test, the officer has reasonable cause to suspect that he/she is over the prescribed limit.

Under s. 6(5)(b) however, an officer may only arrest that person if he/she has failed to provide a breath specimen *and* the officer has reasonable cause to suspect that he/she has alcohol in his/her body. If a driver refuses to provide a breath specimen, an officer will need to show reasonable cause for suspecting the driver to have alcohol in his/her body—otherwise s. 6(5)(b) does not give a power of arrest.

5.5 Evidential Specimens

Sections 7–9 of the Road Traffic Act 1988 govern the procedure for obtaining specimens for analysis, or evidential specimens. Unlike the preliminary breath tests above, specimens taken under this part of the Act are retained for evidential purposes in subsequent hearings. Such specimens will either be by breath samples taken on an approved machine (e.g. the Lion Intoximeter or the Camic Breath Analyser) or by blood/urine samples.

The print outs from 'approved' machines will generally be admissible in evidence under s. 69 of the Police and Criminal Evidence Act 1984. Minor errors in the printing functions of such machines will not necessarily invalidate the evidence produced by the machine, provided it can be shown that the analytical function of the machine was calibrated and working properly (*Reid* v *DPP* [1999] RTR 357).

Many of the difficulties which have been encountered during the police station process have now been addressed by police forces adopting a national pro-forma and strict adherence to those forms will significantly reduce the likelihood of convictions being quashed (see below).

5.5.1 Provision of Specimens for Analysis

Section 7 of the Road Traffic Act 1988 states:

(1) In the course of an investigation into whether a person has committed an offence under section 3A, 4 or 5 of this Act a constable may, subject to the following provisions of this section and section 9 of this Act, require him—
  (a) to provide two specimens of breath for analysis by means of a device of a type approved by the Secretary of State, or
  (b) to provide a specimen of blood or urine for a laboratory test.

(2) A requirement under this section to provide specimens of breath can only be made at a police station.

(3) A requirement under this section to provide a specimen of blood or urine can only be made at a police station or at a hospital; and it cannot be made at a police station unless—
  (a) the constable making the requirement has reasonable cause to believe that for medical reasons a specimen of breath cannot be provided or should not be required, or
  (b) at the time the requirement is made a device or a reliable device of the type mentioned in subsection (1)(a) above is not available at the police station or it is then for any other reason not practicable to use such a device there, or
  (bb) a device of the type mentioned in subsection (1)(a) above has been used at the police station but the constable who required the specimens of breath has reasonable cause to believe that the device has not produced a reliable indication of the proportion of alcohol in the breath of the person concerned, or
  (c) the suspected offence is one under section 3A or 4 of this Act and the constable making the requirement has been advised by a medical practitioner that the condition of the person required to provide the specimen might be due to some drug;

  but may then be made notwithstanding that the person required to provide the specimen has already provided or been required to provide two specimens of breath.

(4) ...

(5) A specimen of urine shall be provided within one hour of the requirement for its provision being made and after the provision of a previous specimen of urine.

(6) ...

(7) A constable must, on requiring any person to provide a specimen in pursuance of this section, warn him that a failure to provide it may render him liable to prosecution.

KEYNOTE

Section 7(3)(bb) was inserted into the 1988 Act by the Criminal Procedure and Investigations Act 1996 to cater for the arrival of the next generation of 'approved devices'.

At the time of writing the new devices that have been approved under s. 7(1) are:

- the CAMIC (Car and Medical Instrument Co.) **Datamaster**
- the Lion **Intoxylizer 6000UK**
- the **EC/IR Intoximeter**.

While the reliability of a reading made by any device in any particular case is always open to challenge by way of admissible evidence, a defendant does not have an automatic right to challenge the secretary of State's approval of devices every time they are used (see *DPP* v *Memery* [2002] *EWHC 1720*).

In the case of a breath specimen it is presumed that the machine used was reliable; if that presumption is challenged by relevant evidence, the magistrates will have to be satisfied that the machine had provided a reading on which they could rely before they make the assumption (*DPP* v *Brown & Teixeira* [2002] *RTR 23; Cracknell* v *Willis* [1987] 3 WLR 1082). This is not however the same as the statutory presumption under s. 15(2) (as to which, see below). Where there are reasons to believe that a type of device is *generally* unreliable (as opposed to the specific device used in a particular case), representations should be made to the Secretary of State (see *R* v *Skegness Magistrates' Court, ex parte Cardy* [1985] RTR 49); it is not open to magistrates to consider whether the type of device should be on the list of 'approved devices' because of some alleged general design flaw (*DPP* v *Brown & Teixeira* above).

The Requirement

A requirement under s. 7 for a breath specimen can only be made at a police station.

That requirement can be made of more than one person in respect of the same vehicle (e.g. if it is believed that one of three defendants was driving the vehicle) (*Pearson* v *Metropolitan Police Commissioner* [1988] RTR 276).

A requirement for blood or urine can only be made at a police station if the conditions under s. 7(3)(a) to (c) are met.

Medical Reasons

For the officer making the requirement to have 'reasonable cause to *believe*' that medical reasons exist, there is no need to seek medical advice first (see *Dempsey* v *Catton* [1986] RTR 194). It is the objective *cause* of that belief which will be considered by the courts, not whether the officer actually did believe that a medical reason existed (*Davis* v *DPP* [1988] RTR 156).

In *Kinsall* v *DPP* (2002) LTL 13 March, the Divisional Court held that, just because the defendant had a mouth spray (for angina) and tablets at the time of arrest, that did not of itself impose an obligation on the relevant police officer to consult a doctor before deciding that any specimen to be provided would be blood. In that case the officer had asked the defendant if there were any medical reasons why blood should not be taken and he had said that there were not.

Where an officer specifically asks whether there are any medical reasons for not taking a blood sample, he/she is entitled to rely on the answer given unless it is obvious that such reasons exist (*Jubb* v *DPP* [2002] EWHC 2317).

If a person is too drunk to provide a breath specimen, that may be regarded as a 'medical' reason for requiring a sample of blood/urine (*Young* v *DPP* [1992] RTR 328).

Device 'Available'?

If the relevant machine at a police station will not calibrate or is in some other way unreliable, this would appear to make it 'unavailable'. In such a case, however, the driver may be taken to another police station where such a machine is 'available'; this may be done *even if the driver has already provided two samples on the inaccurate machine* (*Denny* v *DPP* [1990] RTR 417).

Whether a machine is 'available' if it produces an accurate reading of the analysis of the driver's breath but cannot, for some reason, produce a hard copy printout of that reading is open to some doubt. In one case (*Morgan* v *Lee* [1985] RTR 409) where the officer did not know of the machine's defect, his subsequent request for a blood sample was held to have been unlawful. However, in a later case the Divisional Court held that, if the officer knew at the time that the machine had any form of malfunction, he/she could require a specimen of blood/urine under s. 7(3)(b) (*Thompson* v *Thynne* [1986] Crim LR 629) and this *subjective* approach has been followed and approved many times since.

Where the requirement for blood/urine is made under s. 7(3)(b), that sample must be used and the prosecution cannot revert to evidence produced by the breath sample (*Badkin* v *Chief Constable of South Yorkshire* [1988] RTR 401).

If the option to provide a blood sample is chosen then, again, the driver may be taken to another police station where a doctor is available (*Chief Constable of Kent* v *Berry* [1986] Crim LR 748).

Section 7(3)(b) also allows the officer (see below) the option of a blood/urine specimen when, 'for any other reason' it is not practicable to use a breath-testing device at the police station. This would clearly include the situation where no trained operator is available to work the machine (*Chief Constable of Avon and Somerset* v *Kelliher* [1986] Crim LR 635).

Drugs

In contrast to the condition at s. 7(3)(a), the advice of a medical practitioner *must* be sought before the power to request blood/urine can be made under s. 7(3)(c); even then, it can only be made in cases of being unfit (**see para. 5.2**) or causing death by careless driving when under the influence of drink or drugs (**see para. 2.7**).

When such medical advice is sought, the doctor must give the officer a clear verbal statement to the effect that the driver's condition was due to some drug before the power arises (*Cole* v *DPP* [1988] RTR 224).

Warning

The warning required by s. 7(7) is critical to a successful prosecution for failing to provide a specimen when required under s. 7(3). It is not needed when a driver elects to give an alternative sample under s. 8 below (see *Hayes* v *DPP* [1993] Crim LR 966 and also *DPP* v *Jackson; Stanley* v *DPP* [1998] 3 WLR 514 below).

Choice by Officer

Section 7(4) states:

> (4) If the provision of a specimen other than a specimen of breath may be required in pursuance of this section the question whether it is to be a specimen of blood or a specimen of urine and, in the case of a specimen of blood, the question who is to be asked to take it shall be decided (subject to subsection (4A)) by the constable making the requirement.
>
> (4A) Where a constable decides for the purposes of subsection (4) to require the provision of a specimen of blood, there shall be no requirement to provide such a specimen if—
> (a) the medical practitioner who is asked to take the specimen is of the opinion that, for medical reasons, it cannot or should not be taken; or
> (b) the registered health care professional who is asked to take it is of that opinion and there is no contrary opinion from a medical practitioner;
> and, where by virtue of this subsection there can be no requirement to provide a specimen of blood, the constable may require a specimen of urine instead.

KEYNOTE

In exercising the power under s. 7(3) and s. 8 below, the decision as to whether the specimen will be blood or urine will be made by the officer. The line of decided cases which allowed a 'drivers' preference' was ended by the House of Lords' ruling in *DPP* v *Warren* [1993] AC 319 where their Lordships stated unequivocally that the decision, in both situations, is to be made by the officer.

This area of law has now become quite complex and what follows is a brief summary of the key points arising from the case decisions. In cases of doubt, the advice of the CPS should be sought.

Following the decision in *DPP* v *Warren*, many forces adapted their pro-formas once again in order to close what Lord Bridge called *a variety of wholly unmeritorious avenues of escape from conviction*. For a while it was believed that Lord Bridge's *dicta* in *Warren* would avoid any further appeals based on the procedure followed by police officers when obtaining evidential specimens in line with their respective force forms. However, there have been several more recent cases where the defendants have argued that the *Warren* formula, as laid down by

Lord Bridge, has not been followed. Most of those cases have concerned what information must be given, and when it must be given to a defendant when explaining the choices available under s. 7(4) above and s. 8(2) below. In the first case, *Fraser* v *DPP* [1997] RTR 373, Lord Bingham CJ examined many of the other authorities since *Warren*. In his judgment he endorsed the pro-forma used by the Northumbria police and said that there was a danger of putting *a new and heretical gloss* on the statute itself. His lordship held that there were many things which drivers ought to be told, but they need not be told them all at once. His lordship went on to say that there was no requirement for an officer to explain to a defendant *at the time of making a decision under s. 7(4)* that any blood sample taken would only be so taken by a doctor.

In the second case, *DPP* v *Jackson*; *Stanley* v *DPP* [1998] 3 WLR 514, the House of Lords stressed the distinction between the roles of the police officer and the doctor. The police officer decides *which* evidential sample(s) should be obtained and the medical practitioner decides *on the validity of the reasons put forward by the defendant* as to why a specimen of blood should not be taken. Their lordships held that, with three exceptions, the rules laid down in *Warren* were not mandatory elements of the procedure for taking evidential specimens but were guidelines indicating matters which should be brought to a defendant's attention before he/she exercised any choice that might be available.

The three elements that **are** mandatory are:

- in a s. 7(3) case (see above), the warning required under s. 7(7) above;
- a statement as to why, in a case under s. 7(3), a breath specimen could not be used; and
- a statement, in a case under s. 8(2) below, that the specimen of breath containing the lower proportion of alcohol did not exceed 50 microgrammes in 100 millilitres of breath. In such a case, the police officer should ask whether there are any medical reasons why a particular specimen could not or should not be taken.

Note the alteration in the legislation (s. 7(4A) above) which now permits a registered health care professional (as opposed to a doctor) to take the blood sample where appropriate. The opinion of the health care professional will be sufficient to authorise the taking of a blood sample provided that there is no contrary opinion from a doctor. This does not mean that the health care professional's opinion has to be confirmed by, or routinely referred to a doctor—in fact this would defeat the whole purpose of the changed legislation. A registered healthcare professional is either a registered nurse or a registered member of another health care profession designated for the purpose by the Secretary of State (Road Traffic Act 1988, s. 11(2)).

It would seem from the Divisional Court's decision in *Bobin* v *DPP* [1999] RTR 375 that, as long as the information set out above is provided by a police officer, it does not matter *which* police officer. Therefore, the warning under s. 7(7) might be given by, for instance, the arresting officer or by the custody officer who makes the requirement for the relevant specimen.

As with other decisions involving when and where to give information to a defendant (e.g. cautioning; **see General Police Duties, chapter 2**), there are probably fewer dangers in giving more than is required, sooner than required than vice versa. Many of the cases before and since *Warren* have turned on the fact that drivers claim that they were not given the chance to reveal medical reasons as to why they could not give blood, or that they were not told that it would be a doctor who took the blood rather than a police officer. Although it is now clear from the House of Lords' decisions above that such opportunity and information need not be given out as a general requirement, there seems to be little harm in following an approach that errs on the side of caution. There are not many cases in this area where drunk drivers go free solely as a result of the police officer giving them too many opportunities or too much information!

The officer should explain what the options are and that the decision as to which will be chosen is the officer's; the driver should also be given the opportunity to state any reasons why blood should not be taken so that the officer may take the appropriate advice (*Edge* v *DPP* [1993] RTR 146).

Specimens of Blood taken from People Incapable of Consenting

Section 7A of the Road Traffic Act 1988 states:

(1) A constable may make a request to a medical practitioner for him to take a specimen of blood from a person ('the person concerned') irrespective of whether that person consents if—
  (a) that person is a person from whom the constable would (in the absence of any incapacity of that person and of any objection under section 9) be entitled under section 7 to require the provision of a specimen of blood for a laboratory test;
  (b) it appears to that constable that that person has been involved in an accident that constitutes or is comprised in the matter that is under investigation or the circumstances of that matter;
  (c) it appears to that constable that that person is or may be incapable (whether or not he has purported to do so) of giving a valid consent to the taking of a specimen of blood; and
  (d) it appears to that constable that that person's incapacity is attributable to medical reasons.

(2) A request under this section—
  (a) shall not be made to a medical practitioner who for the time being has any responsibility (apart from the request) for the clinical care of the person concerned; and
  (b) shall not be made to a medical practitioner other than a police medical practitioner unless—
    (i) it is not reasonably practicable for the request to be made to a police medical practitioner; or
    (ii) it is not reasonably practicable for such a medical practitioner (assuming him to be willing to do so) to take the specimen.

(3) It shall be lawful for a medical practitioner to whom a request is made under this section, if he thinks fit—
  (a) to take a specimen of blood from the person concerned irrespective of whether that person consents; and
  (b) to provide the sample to a constable.

KEYNOTE

This addition to the drink driving legislation came about as a result of the procedural difficulties encountered where suspected drivers are injured in an accident and, as a result, are incapable of giving their consent to the taking of a blood sample. In order to prevent such drivers escaping conviction where they would otherwise have had to provide a blood sample, the legislation now allows for the sample to be taken lawfully without consent at the time. Thereafter, the system relies on the driver giving his or her permission for the sample to be tested in a laboratory. It must appear to the constable that the person has been 'involved' in an accident—this is far wider than suspecting that they were driving, attempting to drive or in charge of a vehicle at the time. However, the *reason* for the test will be important in determining the penalty available in the event that the person fails to give their consent (see below).

It must also appear to the officer that the person is or may be *incapable* (as opposed to simply unwilling) of giving valid consent and that this incapacity is attributable to medical reasons (rather than other issues such as language or ethical objections).

As a general rule the doctor approached should be a medical practitioner employed by or contracted to a police force (see s. 7A(7))—e.g. a Force Medical Examiner or Police Surgeon. If it is not reasonably practicable either to make the request to the police doctor, or for him/her to take the sample, another medical practitioner can be approached but not the doctor who has responsibility for the person's clinical care. The doctor is then empowered both to take the sample and to provide it to the officer without the person's consent if the doctor thinks fit—there is, however, no obligation to do so.

Given that most situations where this power will be relevant will concern people in hospital, the provisions covering patients should be considered (**see para. 5.5.4**).

Section 7A of the Road Traffic Act 1988 goes on to state:

(4) If a specimen is taken in pursuance of a request under this section, the specimen shall not be subjected to a laboratory test unless the person from whom it was taken—
  (a) has been informed that it was taken; and

(b) has been required by a constable to give his permission for a laboratory test of the specimen; and
(c) has given his permission.

(5) A constable must, on requiring a person to give his permission for the purposes of this section for a laboratory test of a specimen, warn that person that a failure to give the permission may render him liable to prosecution.

KEYNOTE

The above requirements are critical if the analysis of any sample lawfully taken is to be of any evidential value. By implication, the requirement that the person gives his/her permission to the sample being sent to a laboratory means that he/she should only be informed of the above matters and given the statutory warning once capable of understanding what is going on.

OFFENCE: **Failing to give Permission for Laboratory Test of Specimen—*Road Traffic Act 1988, s. 7A(6)***

- Triable summarily
- If the test was for ascertaining ability to drive or the proportion of alcohol at the time the offender was driving or attempting to drive—six months' imprisonment and/or fine and obligatory disqualification
- In any other case, three months' imprisonment and/or fine and discretionary disqualification

(No specific power of arrest)

The Road Traffic Act 1988, s. 7A states:

(6) A person who, without reasonable excuse, fails to give his permission for a laboratory test of a specimen of blood taken from him under this section is guilty of an offence.

KEYNOTE

For a discussion of what may amount to reasonable excuse, see the following Keynote and paragraph. There should be far fewer opportunities for claiming reasonable excuses here given that there is no technical equipment involved. The permission being given (or not) is for the laboratory analysis, *not* the taking of the specimen. Therefore delaying the giving of permission in order to obtain legal advice may be acceptable, particularly as, once the specimen has been taken and properly stored, there is not the same urgency as there is with the *taking* of a specimen before the alcohol/drugs dissipate. 'Failing' includes refusing (s. 11(2)).

Choice by Defendant

Section 8 of the Road Traffic Act 1988 states:

(1) Subject to subsection (2) below, of any two specimens of breath provided by any person in pursuance of section 7 of this Act that with the lower proportion of alcohol in the breath shall be used and the other shall be disregarded.

(2) If the specimen with the lower proportion of alcohol contains no more than 50 microgrammes of alcohol in 100 millilitres of breath, the person who provided it may claim that it should be replaced by such specimen as may be required under section 7(4) of this Act and, if he then provides such a specimen, neither specimen of breath shall be used.

KEYNOTE

Under s. 8(2) the driver does have a choice; that being to ask for the specimen of breath to be replaced by an option under s. 7(4) above. As discussed, once that request has been made, the choice as to whether it will be blood or urine will be the officer's.

When the driver is told of the option under s. 8(2), he/she *must* be told that the reason for the option is because the breath specimen with the lower proportion of alcohol contains no more than 50 microgrammes. Simply telling the driver that the option exists without giving this reason will amount to one of the fatal procedural failures set out in *Warren* above (*R* v *Bolton Justices, ex parte Zafer Alli Khan* [1999] Crim LR 912).

If a driver elects to give blood and, through no fault of his/her own, the sample then proves unsuitable for evidential use, the original reading from the machine cannot be used in its place (*Archbold* v *Jones* [1985] Crim LR 740).

Although the choice is to be made by the requesting officer, the possibility of medical reasons for not giving blood must still be considered. An alleged fear of needles by the driver (which has provided the courts with a number of opportunities to explore this area) is a relevant consideration when making a decision as to whether a blood sample should be taken (*DPP* v *Jackson*; *Stanley* v *DPP* above and also *Johnson* v *West Yorkshire Metropolitan Police* [1986] RTR 167). The driver should be given the opportunity to state any medical reasons why blood should not be taken but failure to give him/her that opportunity will not necessarily be fatal to an ensuing prosecution (*DPP* v *Orchard* [2000] All ER (D) 1457).

The medical advice may be given to the officer over the telephone if appropriate (*Andrews* v *DPP* [1992] RTR 1).

OFFENCE: **Failing to Provide Evidential Specimen—*Road Traffic Act 1988, s. 7(6)***

- Triable summarily • Six months' imprisonment and/or a fine
- Obligatory disqualification

(No specific power of arrest)

The Road Traffic Act 1988, s. 7 states:

(6) A person who, without reasonable excuse, fails to provide a specimen when required to do so in pursuance of this section is guilty of an offence.

KEYNOTE

Section 11(3) of the Road Traffic Act 1988 states that a breath specimen must be provided in such a way as to enable the analysis to be carried out. If a driver produces it in any other way, they will have 'failed to provide'.

As with preliminary breath tests (**see para. 5.4**), 'fail' will include a refusal. If a driver elects to substitute a breath specimen (which gave a reading of no more than 50 microgrammes) with one of blood/urine under s. 8 (above), and subsequently fails or refuses to do so, he/she does not commit an offence under s. 7(6); you revert to the original breath specimen.

Provision of only one specimen of breath is a 'failure to provide' (*Cracknell* v *Willis* [1987] 3 All ER 801).

The issues arising where a defendant claims to have had a 'reasonable excuse' for failing to provide a specimen of breath were recently reviewed by the Divisional Court in *spp* v *Falzarano* [2001] RTR 14.

In *Falzarano* magistrates had accepted the defendant's claim that she was physically unable to provide the specimen as she was suffering from a panic attack. The defendant proved that she had a history of such attacks and that she was receiving medical treatment for the condition. Despite evidence from her own doctor that the condition and her failure to take her medication on the day in question should not have prevented her from providing a specimen of breath, the magistrates found that she did in fact have a reasonable excuse. On appeal by way of case stated (as to which, **see Evidence and Procedure, chapter 2**), the Divisional Court held that:

- The 'reasonable excuse' had to arise out of a physical or mental inability to provide a specimen or a substantial risk to health in its provision (per *R* v *Lennard* [1973] RTR 252).
- The evidence in support of the 'reasonable excuse' normally had to be medical but the defendant could provide it himself/herself.
- There had to be a causative link between the excuse and the failure to provide a specimen (*DPP* v *Pearman* [1992] RTR 407).
- Being drunk or under stress was not in itself enough to provide a 'reasonable excuse' for failing to provide a specimen.

- Having considered the medical evidence, legal advice from their clerk and the demeanour of the defendant when testifying, the magistrates had been entitled on the evidence to find that Falzarano had a reasonable excuse for failing to provide breath specimens.

This decision does *not* mean that shortness of breath caused by panic attacks or stress will always amount to a reasonable excuse; it does, however, provide a useful review of the law in this area.

Reasonable Excuse

What amounts to a 'reasonable excuse' is a matter of law (see *Falzarano* above); whether the defendant actually *had* such an excuse is a question of fact for the court to determine having regard to the particular circumstances in each case. There have been many cases where 'excuses' have been put forward. The following is a brief summary:

Not reasonable

- Refusal until legal advice has been sought. The incorporation of the European Convention on Human Rights has not altered this position which was established in *DPP* v *Billington* [1988] RTR 231; *Campbell* v *DPP* [2003] Crim LR 118. This was reiterated by the Divisional Court in *Kennedy* v *DPP* [2003] Crim LR 120 where it was held that the public interest required that the obtaining of specimens should not be delayed in this way.
- The absence of an appropriate adult where the defendant is a juvenile (*DPP* v *Evans* [2003] Crim LR 338).
- Refusal until solicitor present.
- Refusal on advice by solicitor.
- Refusal until driver had read the Codes of Practice under PACE.
- Mistaken belief by the defendant.
- Belief that the officer did not have the authority to make the requirement.
- Religious beliefs.
- Self-induced intoxication.
- Mental anguish caused by custody officer's behaviour.

Reasonable

- Mental incapacity.
- Physical incapacity.
- Inability to understand requirement caused by factors other than drink or drugs (e.g. language barrier).

Where any physical or mental incapacity (to provide the specimen or to understand the warning which accompanies it) is put forward, there will need to be clear, independent support for the claim being genuine (e.g. from a medical practitioner).

The courts have warned of the need for caution when accepting a defendant's claim of such incapacity, particularly given the generally stressful nature of the police station procedure, together with the possible effects following an accident or motoring incident leading up to the request.

In *DPP* v *Furby* [2000] RTR 181, it was held that if a police officer required a motorist to provide a breath specimen at the police station and the motorist made no effort at all to blow into the machine, he/she could not subsequently argue that he/she had a reasonable excuse for failing to do so. This principle was re-stated by the Administrative Court in a case where the driver alleged at the roadside that he suffered from bronchitis and therefore could not provide a breath sample. The arresting officer did not tell the custody officer of this alleged condition, neither did the motorist himself. As a result, the Court held that no objective observer could say that the custody officer had 'reasonable cause to believe that for medical reasons a specimen of breath should not be required' and therefore the defendant should have been convicted for failing to provide a specimen (*DPP* v *Lonsdale* [2001] RTR 444).

Where a defendant's mental capacity to understand the warning is impaired by his/her drunkenness, this is not a 'reasonable' excuse (*DPP* v *Beech* [1992] RTR 239).

Blood

Issues regarding the admissibility of blood specimens will differ depending on whether the specimen was taken under s. 7A (without the person's consent) or in any other case.

Taking the general situations first, s. 11(4) of the Road Traffic Act 1988 provides that a person 'provides a specimen of blood' only if he/she consents and the specimen is taken by a medical practitioner or, if at a police station, a medical practitioner of registered health care professional. Section 15 of the Road Traffic Offenders Act 1988 (see below) states that any sample of blood will be disregarded unless it is so taken. Once consent is given, it is for the doctor to say from which part of the body it will be taken. Any insistence on a different course of action will be a refusal (*Rushton* v *Higgins* [1972] RTR 456).

In the case of specimens taken under s. 7A (**see above**), there will clearly be no consent. Section 15(5A) of the Road Traffic Offenders Act 1988 makes separate provision for these cases.

Urine

Urine samples do not need to be taken by a medical practitioner.

The samples will be admissible as long as they are provided within the time set out under s. 7(5) and there are two distinct samples (as opposed to two samples taken during the same act of urinating) (*Prosser* v *Dickeson* [1982] RTR 96).

5.5.2 Use of Specimens

Section 15 of the Road Traffic Offenders Act 1988 states:

(1) This section and section 16 of this Act apply in respect of proceedings for an offence under section 3A, 4 or 5 of the Road Traffic Act 1988 (driving offences connected with drink or drugs); and expressions used in this section and section 16 of this Act have the same meaning as in sections 3A to 10 of that Act.

(2) Evidence of the proportion of alcohol or any drug in a specimen of breath, blood or urine provided by or taken from the accused shall, in all cases (including cases where the specimen was not provided or taken in connection with the alleged offence), be taken into account and, subject to subsection (3) below, it shall be assumed that the proportion of alcohol in the accused's breath, blood or urine at the time of the alleged offence was not less than in the specimen.

(3) That assumption shall not be made if the accused proves—
  (a) that he consumed alcohol before he provided the specimen or had it taken from him and—
    (i) in relation to an offence under section 3A, after the time of the alleged offence, and
    (ii) otherwise, after he had ceased to drive, attempt to drive or be in charge of a vehicle on a road or other public place, and
  (b) that had he not done so the proportion of alcohol in his breath, blood or urine would not have exceeded the prescribed limit and, if it is alleged that he was unfit to drive through drink, would not have been such as to impair his ability to drive properly.

(4) A specimen of blood shall be disregarded unless—
  (a) it was taken from the accused with his consent and either—
    (i) in a police station by a medical practitioner or a registered health care professional; or
    (ii) elsewhere by a medical practitioner; or

(b) it was taken from the accused by a medical practitioner under section 7A of the Road Traffic Act 1988 and the accused subsequently gave his permission for a laboratory test of the specimen.

(5) Where, at the time a specimen of blood or urine was provided by the accused, he asked to be provided with such a specimen, evidence of the proportion of alcohol or any drug found in the specimen is not admissible on behalf of the prosecution unless—

(a) the specimen in which the alcohol or drug was found is one of two parts into which the specimen provided by the accused was divided at the time it was provided, and

(b) the other part was supplied to the accused.

(5A) Where a specimen of blood was taken from the accused under section 7A of the Road Traffic Act 1988, evidence of the proportion of alcohol or any drug found in the specimen is not admissible on behalf of the prosecution unless—

(a) the specimen in which the alcohol or drug was found is one of two parts into which the specimen taken from the accused was divided at the time it was taken; and

(b) any request to be supplied with the other part which was made by the accused at the time when he gave his permission for a laboratory test of the specimen was complied with.

KEYNOTE

Section 15(2) requires the court to:

- take into account evidence of the proportion of alcohol *or any drug* in a specimen of breath, blood or urine *in all cases*; and
- assume that the proportion of alcohol (though apparently not drugs) in the defendant's breath, blood or urine at the time of the alleged offence was *not less* than that found in the specimen, subject to the circumstances in subsection (3).

As such, s. 15 is a 'legislative interference' with the general presumption of innocence in criminal matters and will therefore be viewed with extreme caution by the courts in light of Article 6(2) of the European Convention on Human Rights—**see General Police Duties, chapter 2**). However, the Court of Appeal has held that this section and the 'interference' with the general presumption of innocence by it imposing a persuasive burden on the defendant) is both justified and no greater than necessary (*R* v *Drummond* [2002] RTR 371). (See also *Parker* v *DPP* [2001] RTR 240.)

For the arguments relating to 'reverse' burdens of proof generally, see *R* v *Lambert* [2001] 3 WLR 206.

The presumption in s. 15(2) is specific to offences where alcohol is a constituent of the offence itself and cannot be extended to other driving offences (*R* v *Ash* [1999] RTR 347).

Note that the presumption is not an assumption that the relevant approved device (**see para. 5.5.1**) was working correctly, but an assumption that the proportion of alcohol in the relevant specimen was not less than the proportion of alcohol at the time of the offence (*DPP* v *Brown & Teixeira* [2002] RTR 23).

Taking Evidence into Account

The first requirement, to take the evidence into account in all cases, was accepted by the House of Lords in *Fox* v *Chief Constable of Gwent* [1986] AC 281 as meaning that evidence obtained following an unlawful arrest was still admissible. However, their lordships did not hold this to be an exception to the general law on the admissibility of improperly or unfairly obtained evidence (as to which, **see Evidence and Procedure, chapter 13**). Under the general discretion given to courts by s. 78 of the Police and Criminal Evidence Act 1984, to exclude unfair evidence, evidence of specimens could be so excluded in spite of the provisions of subsection (2) above.

Where a defendant claimed to have asthma and made several unsuccessful attempts to give a specimen of breath at a police station, magistrates were wrong in law to take into account the partial breath test readings regarding a charge of driving whilst unfit through drink (*Willicott* v *DPP* [2001] EWHC Admin 415).

Making the Assumption

Section 15(3) allows the defendant to produce evidence which prevents that presumption from being made. This enables a defendant to prove that the alcohol had been consumed between the time of the alleged offence and the provision of the specimen and, that if the defendant had not consumed it, then he/she would not have been over the limit/unfit. That is, the defendant can prove that the alcohol level at the time of the alleged offence was lower than shown in the analysis. This allows for the 'hip flask defence' which was, for a while, raised frequently by drivers who grabbed a flask from their glove compartment and took a swift drink from it before they were asked to provide a breath sample. The attractiveness of 'post-incident drinking' in order to beat the breathalyser in this way was drastically reduced when the process of 'back calculation' was introduced (see below). Nevertheless, s. 15(3) above allows the defence to show, on the balance of probabilities, that the level of alcohol in the defendant's body at the time of the alleged offence was lower than at the time of the analysis.

In *R* v *Drummond* [2002] RTR 371, it was argued that this contravened the presumption of innocence (under Article 6(2) of the European Convention) because it placed the burden of proof on the defendant to prove that he/she had consumed alcohol after the offence but before providing a specimen. The Court of Appeal did not agree. It held that conviction for the relevant drink driving offences followed an exact scientific test; if the defendant chose to drink after the event, it was he/she who defeated the aim of the legislation by making the test potentially unreliable. Furthermore, the relevant scientific evidence which the defendant could use to counter the specimen result was within his/her control rather than the prosecution's. The burden of proof upon the defence in making such a case would, as ever, be 'on the balance of probabilities' (**see Evidence and Procedure, chapter 11**).

The assumption does not prevent the prosecution from adducing evidence to show that the proportion of alcohol in the defendant's body was in fact *higher* than that in the specimen. To do so, however, requires a process of 'back calculation', a process accepted by the House of Lords in *Gumbley* v *Cunningham* [1989] RTR 49. In that case their lordships said that such a process should be used with caution and evidence of back calculation should not be relied upon unless it is easily understood and clearly persuasive. In a later case Lord Lane CJ said that such calculations should be treated with great care and that they would be less helpful in deciding cases involving driving while over the prescribed limit than they would in offences involving the *standard* of driving (**see chapter 2**) (*R* v *Downes* [1991] RTR 395).

The validity of a back calculation relies on there being an unbroken period leading up to the relevant event (e.g. the driving of a vehicle or the occurrence of an accident). If the prosecution are unable to remove the real possibility that alcohol had been consumed between the relevant event and the specimen being taken, it will not be safe for a court to convict (*R* v *Hodnitt* [2003] EWCA Crim 441).

The effect of subsection (2) is to limit any use of back calculation to the *prosecution*. Other than in the context of subsection (3), a defendant may not adduce back calculation evidence to show that the level of alcohol in his/her body was in fact *below* the amount in the specimen (*Millard* v *DPP* (1990) 91 Cr App R 108).

Specimen of Blood

Under s. 15(4) the prosecution must generally prove that the defendant consented to the taking of blood and that burden will only be discharged if this is proved beyond a reasonable doubt (see *Friel* v *Dickson* [1992] RTR 366). This may generally be proved by documentary evidence signed by a medical practitioner, subject to certain provisos (s. 16(2)). However, note the special provisions that apply where the blood specimen was taken under s. 7A (i.e. without consent following an accident).

Provision of Specimen *to* Defendant

Under s. 15(5) and (5A), the prosecution must show that the defendant, if he/she asked for one, was provided with a specimen divided and supplied as set out in the subsection. Although the specimen must be divided 'at the time' it is taken and as part of the same continuing event, there is no need for it to be done in the defendant's presence (*DPP* v *Elstob* [1992] Crim LR 518). Again, although the *division* of the specimen must be made at the time it is taken, there is no requirement that the defendant be *provided* with his/her part 'at the time'—only that it be provided within a reasonable time thereafter—see *R* v *Sharp* [1968] 2 QB 564. The part-specimen must be in such a quantity and of such a quality that it is capable of being analysed (*Smith* v *Cole* [1971] 1 All ER 200). Whether or not this has been done is a question of fact for the court to determine on its own merits.

The Administrative Court has held however that there is no free-standing right, either under the 1988 Act or at common law, for the defendant to be informed of his/her entitlement to a part of the sample (see *Campbell* v *DPP* [2003] Crim LR 118). However, the Court acknowledged that there might be occasions where the failure to tell a defendant of this entitlement might cause him/her prejudice and thereby allow the admissibility of the sample to be challenged. Therefore, although there is no specific right to be told of this entitlement, it is probably both good sense and good practice to do so.

The Court of Appeal recently refused to extend the requirements of s. 15(5) to an offence of causing death by dangerous driving (as to which, **see chapter 2**) (*R* v *Ash* [1999] RTR 347). In that case a sample of blood was properly taken at a hospital from a driver who had been very seriously injured in a fatal road traffic accident. The driver argued that, as the sample of blood taken from him had not been divided into two parts, evidence of its analysis should not have been allowed at his trial. The Court held that s. 15 applied only to the sections of the Road Traffic Act 1988 that were expressly set out, namely ss. 3A, 4 and 5. This restricted nature of s. 15 meant that it could not be extended to other offences such as causing death by dangerous driving and that, had Parliament intended otherwise, it would have said so. The Court also held that the admission of that evidence would not therefore be 'unfair' and should not be excluded under s. 78 of the Police and Criminal Evidence Act 1984.

Where a police officer supplied the defendant with a part-specimen but did not follow the local procedure of sealing it inside an envelope and subsequently convinced the defendant that the part-specimen could not then be submitted for analysis, the conviction based on the specimen was quashed (*Perry* v *McGovern* [1986] RTR 240). Although a 'principle' of law grew up around this case to the effect that actions by the police must not mislead or deter a defendant from submitting the part-specimen for analysis, the provisions of s. 78 of the Police and Criminal Evidence Act 1984 would appear to cover any such eventualities (**see Evidence and Procedure, chapter 13**).

Documentary Evidence

Section 16 of the Road Traffic Offenders Act 1988 states:

> (1) Evidence of the proportion of alcohol or a drug in a specimen of breath, blood or urine may, subject to subsections (3) and (4) below and to section 15(5) and (5A) of this Act, be given by the production of a document or documents purporting to be whichever of the following is appropriate, that is to say—
> (a) a statement automatically produced by the device by which the proportion of alcohol in a specimen of breath was measured and a certificate signed by a constable (which may but need not be contained in the same document as the statement) that the statement relates to a specimen provided by the accused at the date and time shown in the statement, and
> (b) a certificate signed by an authorised analyst as to the proportion of alcohol or any drug found in a specimen of blood or urine identified in the certificate.

(2) Subject to subsections (3) and (4) below, evidence that a specimen of blood was taken from the accused with his consent by a medical practitioner or a registered health care professional may be given by the production of a document purporting to certify that fact and to be signed by a medical practitioner or a registered health care professional.

(3) Subject to subsection (4) below—

(a) a document purporting to be such a statement or such a certificate (or both such a statement and such a certificate) as is mentioned in subsection (1)(a) above is admissible in evidence on behalf of the prosecution in pursuance of this section only if a copy of it either has been handed to the accused when the document was produced or has been served on him not later than seven days before the hearing, and

(b) any other document is so admissible only if a copy of it has been served on the accused not later than seven days before the hearing.

(4) A document purporting to be a certificate (or so much of a document as purports to be a certificate) is not so admissible if the accused, not later than three days before the hearing or within such further time as the court may in special circumstances allow, has served notice on the prosecutor requiring the attendance at the hearing of the person by whom the document purports to be signed.

KEYNOTE

Section 16 sets out the provisions for documentary evidence to be used when proving the proportion of alcohol *or drugs* in specimens of breath, blood or urine. This has proved to be another fertile ground for litigation and there are many case decisions surrounding the interpretation of the provisions of s. 16. In summary, the section allows for proof of the proportion of alcohol or drugs in a specimen to be made by a statement produced by the approved device that measured the specimen together with a certificate as to its truth signed by a police officer (s. 16(1)(a)). It also provides for evidence to be given in the form of a certificate signed by an authorised analyst as to the proportion of alcohol or drug found in a specimen of blood or urine (s. 16(1)(b)).

Section 16(3) and (4) set out the requirements for service of a copy of any such statement and certificate on the defendant before it can become admissible. Any such copy or other document must be served on the defendant not later than seven days before the hearing (s. 16(3)). Certificates will not be admissible if the defendant serves notice not later than three days before the hearing (or longer if the court allows) on the prosecutor requiring the attendance of the person by whom the certificate purports to be signed (s. 16(4)).

It has been held that there is no requirement for the *copy* to be signed by the police officer (*Chief Constable of Surrey* v *Wickens* [1985] RTR 277).

In *McCormack* v *DPP* [2002] RTR 20, having provided the required specimens of breath, the driver signed the three copies of the printout produced by the intoxilyser machine. However, on being offered a copy of the printout, the driver refused to accept it. On appeal against conviction, the driver queried (among other things) whether s. 16(3)(a) had been complied with by simply offering him the printout, rather than showing that any physical transfer had taken place. The driver argued that the document had not been 'handed' to him as required by s. 16(3)(a) and therefore was not admissible. He also argued that a previous authority on the point (*Walton* v *Rimmer* [1986] RTR 31) was wrong and should not be followed, especially since there was higher authority in Scottish case law to support his argument. The Administrative Court disagreed; there does not have to be proof of any physical transfer of the document to the driver before it becomes admissible here (though that would clearly save any argument). The fact that this situation is applied differently in Scottish cases arising from the same statute did not persuade the Court.

Failing to prove service of the certificate on the defendant within the relevant time period will render that certificate inadmissible and the court will not accept hearsay as an alternative (*Whyte* v *DPP* [2003] EWHC 358).

The precise effect of failing to serve the certificate on the defendant is that the evidence contained within them is inadmissible *to prove the proportion of alcohol* in question. This does not mean that it is *inadmissible* in relation to any other purpose, such as showing why it was felt necessary by the police officer to require a blood sample (see *Jubb* v *DPP* [2002] EWHC 2317). Similar, the failure to serve copies on the defendant does not prevent the relevant witnesses—police officer, analysts or medical practitioners—from providing oral evidence in person.

It would appear that oral evidence as to a defendant's breath-alcohol level may be given by the police officer who operated the machine, provided that evidence also proves that the approved device was properly calibrated (*Owen* v *Chesters* [1985] RTR 191).

5.5.3 Detention of Person Affected

Section 10 of the Road Traffic Act 1988 states:

(1) Subject to subsections (2) and (3) below, a person required to provide a specimen of breath, blood or urine may afterwards be detained at a police station until it appears to the constable that, were that person then driving or attempting to drive a mechanically propelled vehicle on a road, he would not be committing an offence under section 4 or 5 of this Act.

(2) A person shall not be detained in pursuance of this section if it appears to a constable that there is no likelihood of his driving or attempting to drive a mechanically propelled vehicle whilst his ability to drive properly is impaired or whilst the proportion of alcohol in his breath, blood or urine exceeds the prescribed limit.

(3) A constable must consult a medical practitioner on any question arising under this section whether a person's ability to drive properly is or might be impaired through drugs and must act on the medical practitioner's advice.

KEYNOTE

The requirement to consult a doctor under s. 10(3) is mandatory.

There is no guidance as to how the officer might form his/her view as to the fitness of a person to drive when released. A further screening breath test may be used if appropriate but it is not required as it was under the former legislation.

If there is no likelihood of the person driving any mechanically propelled vehicle while impaired or over the limit, the person must be released. Similarly, s. 10 provides no authority to retain that person's property (e.g. car keys) after their release.

5.5.4 Hospital Procedure

Section 9 states:

(1) While a person is at a hospital as a patient he shall not be required to provide a specimen of breath for a breath test or to provide a specimen for a laboratory test unless the medical practitioner in immediate charge of his case has been notified of the proposal to make the requirement; and—
  (a) if the requirement is then made, it shall be for the provision of a specimen at the hospital, but
  (b) if the medical practitioner objects on the ground specified in subsection (2) below, the requirement shall not be made.

(1A) While a person is at a hospital as a patient, no specimen of blood shall be taken from him under section 7A of this Act and he shall not be required to give his permission for a laboratory test of a specimen taken under that section unless the medical practitioner in immediate charge of his case—
  (a) has been notified of the proposal to take the specimen or to make the requirement; and
  (b) has not objected on the ground specified in subsection (2).

(2) The ground on which the medical practitioner may object is—
  (a) in a case falling within subsection (1), that the requirement or the provision of the specimen or (if one is required) the warning required by section 7(7) of this Act would be prejudicial to the proper care and treatment of the patient; and
  (b) in a case falling within subsection (1A), that the taking of the specimen, the requirement or the warning required by section 7A(5) of this Act would be so prejudicial.

KEYNOTE

A 'hospital' will include an institution providing medical treatment for in-patients or out-patients (s. 11(2) of the Road Traffic Act 1988). It will also include anywhere within the precincts of that hospital.

Whether the person is there as a 'patient' or in another capacity will be a question of fact. In assessing this fact the courts will be helped by reference to hospital records which show names of patients, together with their times of admission and discharge (see *Askew* v *DPP* [1988] RTR 303), although these records will not be conclusive.

If a person has been treated and then discharged from the hospital, he/she ceases to be a 'patient' for these purposes even if he/she has to return at a future date for further, related treatment (e.g. to have stitches removed).

Section 9(1A) is inserted by the Police Reform Act 2002 s. 56 in order to deal with the situations where a doctor has taken a blood specimen from the patient without his/her consent (under s. 7A, **see para. 5.5.1**). The overall effects of s. 7A and the above provision is that both the doctor taking the sample and the doctor in immediate charge of the patient need to be two separate people, both of whom will have to agree to the taking to the specimen.

A useful case illustrating this area is *Webber* v *DPP* [1998] RTR 111n. In *Webber* the driver of a vehicle involved in an accident was taken to a hospital where she was requested to provide a specimen of breath under s. 6. She refused. While she was still a patient at the hospital, the driver was required to provide a blood specimen for analysis under ss. 7(1)(b) and 9(1). She agreed but, before she was able to provide the specimen, she was discharged. She was then arrested under s. 6(5)(b) as a result of her earlier refusal. The driver was taken to a police station where she provided a specimen of blood on the strength of which she was subsequently convicted of driving while over the prescribed limit. She appealed, by way of case stated, on the ground that s. 9(1)(a) required the blood specimen to be taken *at the hospital*, which in this case it was not. The Divisional Court held that, once the obligation to provide a specimen for analysis has been made, it is not to be discharged simply by an 'irrelevant change of *locus*'. The court went on to say, by making the requirement, the police officer set in train a procedure which carries a sanction (s. 7(6)), a procedure that was not to be altered by the mere fact that the defendant, for whatever reason, had left the hospital before complying with the requirement.

The restriction applies to both preliminary breath tests and evidential tests.

If the patient provides a positive reading or fails to provide a specimen of breath, he/she cannot then be arrested while still a 'patient' (s. 6(5)).

The procedure for requiring a specimen for analysis is generally the same as that employed when requiring blood/urine at a police station, with the exception that the officer requiring the sample ought to inform the patient that a breath specimen cannot be taken at a hospital (*Ogburn* v *DPP* [1994] RTR 241).

6 Insurance

6.1 Introduction

People using or driving vehicles on roads and public places necessarily present an element of risk to others. In addition to the many provisions which minimise that risk so far as is practicable there is a requirement that most road users are insured against third party risk.

Checking the insurance cover of drivers and owners of vehicles quickly has always been a fairly unreliable business. The government, at the time of writing, was introducing a Motor Insurance Database which will enable the police (under the Vehicles (Crime) Act 2001) to access drivers' insurance details when dealing with motorists.

6.2 Requirement for Insurance

Section 143(1) of the Road Traffic Act 1988 states:

(1) Subject to the provisions of this Part of this Act—

(a) a person must not use a motor vehicle on a road or other public place unless there is in force in relation to the use of the vehicle by that person such a policy of insurance or such a security in respect of third party risks as complies with the requirements of this Part of this Act, and

(b) a person must not cause or permit any other person to use a motor vehicle on a road or other public place unless there is in force in relation to the use of the vehicle by that other person such a policy of insurance or such a security in respect of third party risks as complies with the requirements of this Part of this Act.

OFFENCE: **Contravening Requirement for Insurance—*Road Traffic Act 1988, s. 143(2)***

• Triable summarily • Fine • Discretionary disqualification

(No specific power of arrest)

The Road Traffic Act 1988, s. 143 states:

(2) If a person acts in contravention of subsection (1) above he is guilty of an offence.

KEYNOTE

The requirement to have insurance was extended to public places as well as roads by the Motor Vehicles (Compulsory Insurance) Regulations 2000 (SI 2000 No. 726). This change came about as a result of the House of Lords' decision in *Cutter* v *Eagle Star Insurance Co. Ltd* [1998] 4 All ER 417 as to the definition of a road (see chapter 1).

The policy behind s. 143 is to safeguard road users and pedestrians from uninsured injury from a mechanically powered vehicle, by providing for compulsory insurance (*Winter* v *DPP* [2003] RTR 14). Therefore in deciding whether a particular conveyance or contraption needs insurance or not, the court will have this feature firmly in mind.

Unless the insurer of a vehicle delivers a certificate of insurance to the person taking out the policy, the requirements of s. 143(1) will not have been met (s. 147(1)). Therefore, anyone using, or causing or permitting to be used, a motor vehicle on a road or other public place before the delivery of such a certificate, commits this offence.

A driver may be required under s. 165 (**see para. 6.2.6**) or s. 170 (**see para. 4.2**) of the Road Traffic Act 1988 to produce a certificate of insurance (or other acceptable form of security). In such cases, in order to determine whether a motor vehicle was being driven in contravention of s. 143 above, s. 171 of the Road Traffic Act 1988 places a requirement on the *owner* of the vehicle to give such information as required by the police. Failure to comply with such a requirement is a summary offence.

Many insurance companies appear now to issue instant insurance cover over the telephone by creating an 'agency' to which the insurance certificate is delivered on behalf of the insured.

For the difference between 'using', 'causing' and 'permitting' and the proof required for each, **see chapter 1**. Note that if the driver is employed by the vehicle's owner you must prove that the driver was acting in the course of his/her employment before you can convict the owner of using, causing or permitting the use of the vehicle without insurance. The driver's own inadmissible statement to that effect will not suffice (*Jones* v *DPP* [1999] RTR 1).

The risks which must be covered by any insurance policy for the purposes of s. 143 are set out in s. 145(3). These are generally:

- Liability in relation to bodily injury caused to others when the vehicle on a road in Great Britain.
- Civil liability in relation to the use of a vehicle from another EU Member State in Great Britain.
- Civil liability in relation to the use of a vehicle from Great Britain in another EU Member State.

A local authority has the power (under s. 222 of the Local Government Act 1972) to bring a criminal prosecution for the offence of driving without insurance contrary to s. 143 of the Road Traffic Act 1988 (*Middlesbrough Borough Council* v *Safeer* [2002] RTR 3). (For a list of other offences for which a local authority may prosecute, see s. 4 of the Road Traffic Offenders Act 1988.)

6.2.1 Absolute Liability

Generally the offence under s. 143 is one of absolute liability (**see Crime, chapter 1**), that is, you need not prove any intent or guilty knowledge by the defendant in order to convict (*Tapsell* v *Maslen* [1967] Crim LR 53).

If, however, a person allows another to use their vehicle on the express condition that the other person insures it first, the lender cannot be convicted of 'permitting' (*Newbury* v *Davis* [1974] RTR 367).

6.2.2 Defence

The Road Traffic Act 1988, s. 143 states:

(3) A person charged with using a motor vehicle in contravention of this section shall not be convicted if he proves—
  (a) that the vehicle did not belong to him and was not in his possession under a contract of hiring or of loan,
  (b) that he was using the vehicle in the course of his employment, and
  (c) that he neither knew nor had reason to believe that there was not in force in relation to the vehicle such a policy of insurance or security as is mentioned in subsection (1) above.

KEYNOTE

Note that s. 143(3) provides a special defence for employees using their employer's vehicle in the course of their employment.

The burden of proof in such a case is on the defendant and will be judged on the balance of probabilities (*R* v *Carr-Briant* [1943] 2 All ER 156).

6.2.3 Check 'This'

As with other areas of law involving documentation (e.g. firearms (**see General Police Duties, chapter 5**)), when considering vehicle insurance it is useful to ask:

Does

- **This** policy cover
- **This** person to drive/use
- **This** vehicle for
- **This** purpose on
- **This** day?

Each of these elements is considered in turn.

This Policy

The form in which certificates must appear, together with the requirement for insurers to keep and supply copies of records are contained in the Motor Vehicles (Third Party Risks) Regulations 1972 (SI 1972/1217) as amended. (The latest amendment to these Regulations can be found in the Motor Vehicles (Third Party Risks) (Amendment) Regulations 1999 (SI 1999/2392).)

Cover notes are included, by s. 161(1) of the Road Traffic Act 1988, in the meaning of 'policy of insurance'.

The internationally recognised 'green card scheme' allows for the insurance of British vehicles abroad and overseas vehicles in Great Britain (see the Motor Vehicles (International Motor Insurance Card) Regulations 1971 (SI 1971/792)).

There is a requirement (under s. 145(3)(b) of the 1988 Act) however, for insurance policies to provide cover which extends to other European Union countries, removing the need for the 'green card' scheme within the EU by vehicles from Member States.

This Person

If a policy is restricted to a named person, only that person will generally be covered by it.

If, as often happens, the policy covers any person who holds a current driving licence, that description may include the holder of a provisional licence (see *Rendlesham* v *Dunne* [1964] 1 Lloyd's Rep 192) or even the holder of a licence issued in another country.

If a policy refers to 'agents' or 'employees', you will need to establish whether or not the relevant person fits into those categories. Garage proprietors returning vehicles to the owner after repairing them are not in the owner's employment (*Lyons* v *May* [1948] 2 All ER 1062). However, an employee driving an employer's vehicle in an unauthorised manner will not negate the effect of a policy of insurance (*Marsh* v *Moores* [1949] 2 All ER 27).

Generally, an insurance policy obtained by false representations will be valid for the purposes of s. 143 and will remain so until the contract has been 'avoided' (ended) by the insurer (*Durrant* v *MacLaren* [1956] 2 Lloyd's Rep 70).

This Vehicle

Generally the question of whether or not a policy applies to the particular vehicle will not present a problem. Trailers can create difficulties, both in relation to the vehicles covered by the policy, and the use to which they are being put. Trailers are themselves 'vehicles' (**see para. 1.2.5**) at times and may be included in the definition of motor vehicle when charging an offence under s. 143 (see *Rogerson* v *Stephens* [1950] 2 All ER 144).

When motor vehicles are being towed by other vehicles, they remain 'motor vehicles' and, as such, require insurance when used on roads (*Milstead* v *Sexton* [1964] Crim LR 474) and also public places.

This Purpose

Insurance policies will often exclude certain uses (e.g. racing and time trials); many stipulate 'social and domestic purposes'. Lending vehicles to friends in return for payment reimbursing petrol costs will amount to such use, as will using the vehicle to help a friend move house (see *Lee* v *Poole* [1954] Crim LR 942).

Such an expression does not include business trips (see *Wood* v *General Accident etc. Assurance Co.* (1948) 65 TLR 53).

Insurance cover in respect of 'the insured's business' does not extend to other businesses of friends or colleagues (see *Passmore* v *Vulcan etc. Insurance Co.* (1935) 154 LT 258).

If employees deviate from the ordinary course of their duties or employment, on what is often termed a 'frolic of their own', they may not be covered by the terms of the employer's policy. Simply taking a two-mile detour in order to give someone a lift has been held not to invalidate the employer's insurance policy (*Ballance* v *Brown* [1955] Crim LR 384) but each case will have to be decided on its own facts.

This Day

Clearly any insurance policy must be shown to have been in force at the relevant time. Many business users have arrangements whereby they pass the risk to their insurers at short notice and it will always be necessary to establish whether a particular policy is actually operative at the time (see e.g. *Samuelson* v *National Insurance etc. Ltd* [1986] 3 All ER 417).

In a Scottish case a defendant telephoned his insurer and arranged cover to start at a particular time once he had paid the appropriate premium. Three hours before he paid the premium he was stopped by the police. Although his insurers later issued a certificate covering the time at which he was stopped, the defendant was held not to have been covered by the policy *at that time* as he had not yet paid the required premium (*McCulloch* v *Heywood* 1995 SLT 1009).

Although many policies exclude use of the particular vehicle for payment or reward, s. 150 of the 1988 Act makes provision for 'car-sharing' agreements, provided that they meet the criteria set out.

6.2.4 Exclusions

Section 148(1) of the Road Traffic Act 1988 makes the effects of some restrictions in a policy void in relation to s. 143. This means that, if a policy purports to restrict the extent of its cover by reference to any of these features, breach of them by the insured person will not affect the validity of that policy for the purposes of s. 143.

Those features (under s. 148) are:

(2) ...
- (a) the age or physical or mental condition of persons driving the vehicle,
- (b) the condition of the vehicle,
- (c) the number of persons that the vehicle carries,
- (d) the weight or physical characteristics of the goods that the vehicle carries,
- (e) the time at which or the areas within which the vehicle is used,
- (f) the horsepower or cylinder capacity or value of the vehicle,
- (g) the carrying on the vehicle of any particular apparatus, or
- (h) the carrying on the vehicle of any particular means of identification other than any means of identification required to be carried by or under the Vehicle Excise and Registration Act 1994.

6.2.5 Exemptions

Crown vehicles do not appear to require insurance *while being used as such*; if they are being used for some other purpose, they will need insurance on a road (see the Scottish case of *Salt* v *MacKnight* 1947 SLT 32) or public place.

Section 144 of the 1988 Act sets out other occasions where vehicles will be exempt. Such occasions include police authority vehicles and vehicles being used for police purposes; this will include an off duty police officer using his/her own vehicle for police purposes (*Jones* v *Chief Constable of Bedfordshire* [1987] RTR 332).

By virtue of s. 144(2)(ba), the exemption will extend to vehicles owned by the National Criminal Intelligence Service (NCIS) and the National Crime Squad at a time when they are being used by those organisations by the appropriate officers and employees (as to which, **see General Police Duties, chapter 1**).

6.2.6 Producing Documents

Section 165 of the Road Traffic Act 1988 states:

(1) Any of the following persons—
 (a) a person driving a motor vehicle (other than an invalid carriage) on a road, or
 (b) a person whom a constable or vehicle examiner has reasonable cause to believe to have been the driver of a motor vehicle (other than an invalid carriage) at a time when an accident occurred owing to its presence on a road or other public place, or
 (c) a person whom a constable or vehicle examiner has reasonable cause to believe to have committed an offence in relation to the use on a road of a motor vehicle (other than an invalid carriage),

must, on being so required by a constable or vehicle examiner, give his name and address and the name and address of the owner of the vehicle and produce the following documents for examination.

(2) Those documents are—
 (a) the relevant certificate of insurance or certificate of security (within the meaning of Part VI of this Act), or such other evidence that the vehicle is not or was not being driven in contravention of section 143 of this Act as may be prescribed by regulations made by the Secretary of State,
 (b) in relation to a vehicle to which section 47 of this Act applies, a test certificate issued in respect of the vehicle as mentioned in subsection (1) of that section, and
 (c) in relation to a goods vehicle the use of which on a road without a plating certificate or goods vehicle test certificate is an offence under section 53(1) or (2) of this Act, any such certificate issued in respect of that vehicle or any trailer drawn by it.

(3) Subject to subsection (4) below, a person who fails to comply with a requirement under subsection (1) above is guilty of an offence.

(4) A person shall not be convicted of an offence under subsection (3) above by reason only of failure to produce any certificate or other evidence... if in proceedings against him for the offence he shows that—
 (a) within seven days after the date on which the production of the certificate or other evidence was required it was produced at a police station that was specified by him at the time when its production was required, or
 (b) it was produced there as soon as was reasonably practicable, or
 (c) it was not reasonably practicable for it to be produced there before the day on which the proceedings were commenced,

and for the purposes of this subsection the laying of the information... shall be treated as the commencement of the proceedings.

(5) A person—
 (a) who supervises the holder of a provisional licence granted under Part III of this Act while the holder is driving on a road a motor vehicle (other than an invalid carriage), or
 (b) whom a constable or vehicle examiner has reasonable cause to believe was supervising the holder of such a licence while driving, at a time when an accident occurred owing to the presence of the vehicle on a road or at a time when an offence is suspected of having been committed by the holder of the provisional licence in relation to the use of the vehicle on a road,

must, on being so required by a constable or vehicle examiner, give his name and address and the name and address of the owner of the vehicle.

KEYNOTE

Production of the certificate need not be in person but the certificate must be shown to the officer for long enough to allow proper inspection of it (applying the 'This' rules above); waving it in front of an officer or snatching it back shortly afterwards will not be enough to comply with s. 165 (see *Tremelling* v *Martin* [1971] RTR 196).

In cases where a requirement has been made under s. 165, s. 171 of the Road Traffic Act 1988 places a requirement on the *owner* of the vehicle to give such information as required by the police in order to determine whether the motor vehicle was being driven without insurance. Failure to comply with such a requirement is a summary offence.

The requirement under s. 165(1)(b) has been extended to 'public places' as well as roads by the Motor Vehicles (Compulsory Insurance) Regulations 2000 (SI 2000 No. 726).

For additional requirements following an accident, **see chapter 4**.

For the obligations under s. 165 on the supervision of learner drivers, **see chapter 11**.

Traffic wardens may also demand a person's name and address under s. 165. They may *not*, however, demand driving documents under these particular circumstances (see Functions of Traffic Wardens Order 1970 (SI 1970 No. 1958) and also **para. 10.4.2**).

6.3 Motor Insurers' Bureau

All insurers in Great Britain are required to be members of the Motor Insurers' Bureau (MIB) (s. 145 of the Road Traffic Act 1988).

The purpose of the MIB is to provide compensation where someone is unable to pursue a valid claim against another following a road traffic accident because the other party is:

- not insured
- not known/traceable
- insured by a company now in liquidation.

The MIB has drawn up an agreement with the Secretary of State which sets out its terms of operation. Generally it will not pay compensation to those who are victims of deliberate criminal acts involving motor vehicles (including those who allow themselves to be carried in vehicles taken without the owner's consent), neither will it compensate those who 'use' vehicles without insurance ('use' here will be taken in the wide sense discussed in **chapter 1** and will include some passengers (see e.g. *Stinton* v *Stinton* [1995] RTR 167 and *O'Mahoney* v *Joliffe and Motor Insurers' Bureau* [1999] RTR 245)). The Court of Appeal decided recently that a person injured by an uninsured or untraced driver cannot enforce a claim to compensation against the MIB by citing the 'direct effect' of EC Council Directive 84/5 which provides certain rights for citizens of EU Member States (*Mighell* v *Reading*; *Evans* v *Motor Insurers' Bureau*; *White* v *White* [1999] Lloyd's Rep IR 30).

The MIB scheme generates a considerable amount of civil litigation that, on the whole, is of little direct importance to police officers. However, there are key evidential areas that may arise during the course of an investigation into accidents and collisions. For instance, the MIB will generally deny liability for any damages or loss to passengers who knew, or ought to have known, that the driver was uninsured. Evidence of conversations between passengers and drivers before and during any journeys can therefore become highly relevant in subsequent litigation (see for example *Akers* v *Motor Insurance Bureau* [2003] EWCA Civ 18).

In order to help people seeking compensation generally from motor accidents across the European Economic Area, Member States have set up clearing houses where insurance information relating to vehicles within that country are pooled. In the United Kingdom a company called the Motor Insurers' Information Centre (MIIC) has been set up for this purpose. Following an accident, a person in the relevant Member State will be entitled to specific information regarding the vehicle(s) and insurance policy details on application. Full details of the scheme, along with the entitlement to compensation and the role of the MIB within the scheme can be found in the Motor Vehicles (Compulsory Insurance) (Information Centre and Compensation Body) Regulations 2003 (SI 2003/37).

7 Safety measures

7.1 Introduction

This chapter deals with the main areas of legislation aimed at increasing—and enforcing—safety in the use of vehicles.

7.2 Seat Belts

The law governing the fitting of seat belts to vehicles comes under the Road Vehicles (Construction and Use) Regulations 1986 (SI 1986/1078) (**see chapter 9**); that which regulates the *wearing* of seat belts is provided by the Motor Vehicles (Wearing of Seat Belts) Regulations 1993) (SI 1993/176) for people over 14 years old, and by s. 15 of the Road Traffic Act 1988 for children under that age.

The regulations are made under s. 14 of the Road Traffic Act 1988.

OFFENCE: **Contravention of Regulation Relating to Seat Belts—*Road Traffic Act 1988, s. 14(3)***

- Triable summarily • Fine

(No specific power of arrest)

It is an offence to:

- *drive* a **motor vehicle** or
- *ride* in a *front seat* of a **motor vehicle** or
- *ride* in the *rear seat* of a **motor car** or **passenger car**

in each case without wearing an adult seat belt.

Section 14(3) creates these offences by virtue of reg. 5 of the 1993 Regulations.

The above offences can only be committed by people aged 14 and over.

In these and the offences which follow, it is important to note the different definitions used (e.g. *drive, ride, motor vehicle, passenger car, rear seat* etc.).

The definitions can be found in the Road Traffic Act 1988, ss. 15(9) and 185; the Motor Vehicles (Wearing of Seat Belts) Regulations 1993, reg. 3; the Motor Vehicles (Wearing of Seat Belts by Children in Front Seats) Regulations 1993 (SI 1993/31), regs 2 and 3 and the Road Vehicles (Construction and Use) Regulations 1986 (**see chapter 9**).

Section 14(3) goes on to say that, notwithstanding any enactment or rule of law, no person *other than the person actually committing the contravention* is guilty of an offence by reason of the contravention. This means that, irrespective of the general law relating to the aiding and abetting of offences (**see Crime, chapter 2**), the driver of a vehicle will not be responsible for a passenger not wearing a seat belt. This is similar to the wording used at s. 16 of the 1988 Act (**see para. 7.3**) in relation to motor cycle crash helmets.

This should be contrasted with the position under s. 15 of the 1988 Act in respect of children under 14 years of age who are not wearing seat belts (see below). In those cases the driver *is* responsible and, in theory at least, it would be possible for the child to be guilty of aiding and abetting the driver by not wearing a seat belt. It is also useful to contrast these provisions with those relating to the wearing of motor cycle helmets (**see para. 7.3**).

7.2.1 Exemptions

There are several general exemptions to the requirements of the regulations, many of them following common sense. Drivers of delivery vans, prisoner escorts, taxis and people reversing are all subject *to some degree* of exemption under reg. 6 of the Motor Vehicles (Wearing of Seat Belts) Regulations 1993, SI 1993/176, as are vehicles being used for police or fire brigade purposes. An exemption may give a specific description, not only of the activity (e.g. taxis on hire business), but also of the *vehicle* to which it will apply (for instance, making deliveries in the family car would not attract the exemption because the vehicle used by delivery drivers must *be made or adapted* for making deliveries in order to be exempt).

In addition to these general exemptions, there are other occasions where the requirements will not apply. If there is no adult seat belt provided for the driver, or available for passengers aged 14 and over, the requirements will not apply (reg. 6).

Schedule 2 to the Regulations expands on a number of occasions where seat belts will be deemed not to be 'available'. These occasions include:

- where the seat is properly occupied by someone wearing the seat belt;
- where the seat is being used for a carry cot containing a child under one year old which cannot go anywhere else in the vehicle;
- where the seat is occupied by someone with a medical exemption.

If a person is unable to wear a seat belt owing to a disability, the seat belt will not be regarded for this purpose as being 'available'.

If a person holds a medical certificate signed by a doctor stating that the wearing of a seat belt by that person is inadvisable on medical grounds, that person will be exempt. Any such certificate must state the period over which it applies and carry the symbol prescribed in the regulations. If it is to be used in evidence in answer to a charge under s. 14(3) of the 1988 Act, such a certificate must be produced to a constable on request or to a police station within seven days (s. 14(4)).

7.2.2 Children Under 14

The law in relation to the wearing of seat belts by children is contained in the Road Traffic Act 1988. The fact that it appears in primary legislation (**see Evidence and Procedure, chapter 1**) perhaps gives an indication of the added importance which Parliament has ascribed to this area of safety.

Further evidence of this importance can be seen in the fact that, unlike the foregoing offences, it is the *driver* who carries most responsibility in relation to the wearing of seat belts by children.

Section 15(1) of the Road Traffic Act 1988 states that, where a child under the age of 14 is in the front of a motor vehicle, a person must not without reasonable excuse drive the motor vehicle on a road unless the child is wearing a seat belt in conformity with the regulations.

Section 15 states:

(3) Except as provided by regulations, where a child under the age of fourteen years is in the rear of a motor vehicle and any seat belt is fitted in the rear of that vehicle, a person must not without

reasonable excuse drive the vehicle on a road unless the child is wearing a seat belt in conformity with regulations.

(3A) Except as provided by regulations, where—

(a) a child who is under the age of 12 years and less than 150 centimetres in height is in the rear of a passenger car.

(b) no seat belt is fitted in the rear of the passenger car, and

(c) a seat in the front of the passenger car is provided with a seat belt but is not occupied by any person,

a person must not without reasonable excuse drive the passenger car on a road.

OFFENCE: **Driving Motor Vehicle in Contravention of Requirement for Seat Belts for Under 14's—*Road Traffic Act 1988, s. 15(2) and 15(4)***

- Triable summarily • Fine

(No specific power of arrest)

The Road Traffic Act 1988, s. 15 states:

(2) It is an offence for a person to drive a motor vehicle in contravention of subsection (1) above

...

(4) It is an offence for a person to drive a motor vehicle in contravention of subsections (3) and (3A) above.

KEYNOTE

If all the seats in a vehicle are properly occupied, there is no general requirement to make a child wear a seat belt or to move an adult out of a seat in order to give the child access to a seat belt. There is also no general requirement to move a child from the back seat of a vehicle where there are no seat belts, into the front (or *vice versa*). However, s. 15(3A) creates additional protection for children who are both under 12 and less than 150 cms tall travelling in vehicles which have no rear seat belts fitted but where a front seat with a seat belt is unoccupied. References are to children *in the rear* of the vehicle, not to the *seating*. Therefore, unlike the situation where adults are concerned, it would appear that children under 14 who are standing up in the rear of vehicles are subject to the requirements, irrespective of the seating provided.

The suitability of seat belts, types of seat belts and the manner in which they must be worn by children are set out in the Regulations, as are the further provisions which distinguish between smaller and larger children using them.

Vehicles which do not fall within the definition of a 'motor car' under s. 185 of the 1988 Act (see chapter 1) are exempt from the requirements of s. 15(3) and (3A) above (reg. 9).

Many of the exemptions above, including the exemption in relation to medical certificates above, also apply to children (s. 15(6)).

There are separate requirements which govern the wearing of seat belts in the *front* seats of motor vehicles and they are, unsurprisingly, contained in the Motor Vehicles (Wearing of Seat Belts by Children in Front Seats) Regulations 1993 (SI 1993/31) which set out further conditions. In essence the Regulations require any child under 14 to wear a seat belt or other restraint when *in the front of a vehicle*— the requirement is not restricted to *seating*, and would again appear to apply even if the child were standing up.

If a person drives a motor vehicle in contravention of those Regulations, he/she commits the offence under s. 15(2) above.

Note that these offences have a defence of 'reasonable excuse' which is not available in relation to seat belt offences involving adults. Whether an excuse is reasonable will be a matter for the court to determine in each case.

The Motor Vehicles (Safety Equipment for Children) Act 1991 introduced a new summary offence into s. 15A of the Road Traffic Act 1988. Under s. 15A a person commits an offence by selling or letting on hire carry cot restraining devices and other related safety equipment which is not of the type prescribed or which contravenes the Regulations.

7.3 Motor Cycle Helmets

Under s. 16 of the Road Traffic Act 1988, the Secretary of State may make Regulations relating to the wearing of protective headgear by people riding motor cycles. The relevant regulations are the Motor Cycles (Protective Helmets) Regulations 1998 (SI 1998/1807) as amended.

The Regulations require every person driving or riding on a motor *bicycle* on a road to wear protective headgear (reg. 4).

OFFENCE: **Driving or Riding on Motor Cycle in Contravention of Regulations—*Road Traffic Act 1988, s. 16(4)***

- Triable summarily • Fine

(No specific power of arrest)

The Road Traffic Act 1988, s. 16 states:

> (4) A person who drives or rides on a motor cycle in contravention of regulations under this section is guilty of an offence...

KEYNOTE

The Regulations do not apply to all motor cycles; they only apply to motor bicycles. A motor *bicycle* is a two-wheeled motor cycle, whether having a side-car attached thereto or not. In counting the wheels, any wheels the centre of which in contact with the road surface are less than 460 mm apart are to be treated as one wheel.

Certain motor mowers are exempted by reg. 4(2) of the 1998 Regulations.

The helmet worn must either conform to one of the British Standards specified (in reg. 5) and be marked as such or they must give a similar (or greater) degree of protection as one which meets those Standards *and* be of a type manufactured for motor cyclists (reg. 4(3)(a)). If the helmet is unfastened or improperly fastened (e.g. with part of the chinstrap undone) the offence will be complete (reg. 4(3)(b) and (c)).

It is a summary offence under s. 17(2) to sell or let on hire a helmet for these purposes which does not meet the prescribed requirements. That offence will be committed even if the helmet is sold for off-road use (*Losexis Ltd* v *Clarke* [1984] RTR 174).

The wording of the Regulations suggests that no helmet is required by someone who is pushing a motor bicycle along but, if they straddled it and pedalled it along with their feet, they would require one (see *Crank* v *Brooks* [1980] RTR 441 and **chapter 1**).

Section 16(2) creates an exemption for 'a follower of the Sikh religion while he is wearing a turban'.

The Motor Cycles (Eye Protectors) Regulations 1999 (SI 1999/535) as amended create an offence (under s. 18(3)) of using non-prescribed eye protectors when driving or riding on a motor cycle. These Regulations do not require the use of eye protectors; they impose standards on those appliances which motor cyclists choose to use. The definition of eye protector includes any appliance placed on the head intended for the protection of the eyes (which might include sunglasses). It is also an offence (s. 18(4)) to sell or let on hire eye protectors which do not meet the prescribed requirements.

7.3.1 Passengers

'Riding on' in the above offence means that pillion passengers must wear helmets but s. 16(1) of the 1988 Act exempts people in side-cars.

As with the seat belt provisions above, there is an exception to the general rule on aiding and abetting. Section 16(4) provides that only the person committing the offence of not wearing a helmet shall be liable *unless the person is under 16*. This means that, where a 16 year old rider of a motor bicycle carries a 15 year old pillion passenger who is not wearing a helmet, the 16 year old will be responsible for the commission of the offence by the passenger as well as the passenger themselves. If the passenger were 16 years old, they alone would be responsible for their offence.

OFFENCE: **Passengers on Motor Cycles—*Road Traffic Act 1988, s. 23***

- Triable summarily • Fine • Discretionary disqualification

(No specific power of arrest)

The Road Traffic Act 1988, s. 23 states:

(1) Not more than one person in addition to the driver may be carried on a motor bicycle.

(2) No person in addition to the driver may be carried on a motor bicycle otherwise than sitting astride the motor cycle and on a proper seat securely fixed to the motor cycle behind the driver's seat.

(3) If a person is carried on a motor cycle in contravention of this section, the driver of the motor cycle is guilty of an offence.

KEYNOTE

Under this section it is the *driver* who commits the offence (s. 23(3)), and the passenger can be convicted of aiding and abetting.

There is no specific requirement for the passenger to face the *front*—though this is presumably because it did not occur to the legislators that anyone would be daft enough to face the other way. Any person travelling as a passenger astride a motor cycle but facing the rear may commit an offence under the Road Vehicles (Construction and Use) Regulations 1986 (**see chapter 9**); he/she may also commit an offence of aiding and abetting the driver to drive dangerously (**see chapter 2**).

Note that under the 1986 Regulations suitable supports for a passenger must be provided on a motor bicycle if it is to carry a passenger (reg. 102). There is no requirement that the supports be used by a passenger.

Note also the restrictions on learner drivers carrying passengers (**see chapter 11**).

7.4 Speed Limits

Speed limits may apply to particular roads (e.g. 'restricted' roads), particular vehicles (e.g. heavy commercial vehicles) or temporary conditions imposed in a given area. There are also further conditions regulating speed limits on motorways. As a general rule, the limit on most motorways and dual carriageways for most cars will be 70 mph. However, there are many exceptions and local regulations and orders should be consulted in cases of doubt.

Generally, the approval of the Secretary of State is required before speed limits of any great duration are imposed (see sch. 9 to the Road Traffic Regulation Act 1984). However, that requirement has been removed in relation to speed limits of 20 mph by the Road Traffic Regulation Act 1984 (Amendment) Order 1999 (SI 1999/1608). As a result of this and other legislation (including the Traffic Signs General Directions 2002 (SI 2002/3113)), local traffic authorities can create 20 mph zones under certain circumstances within their jurisdiction as part of a traffic calming scheme. Full guidance in the creation of such zones can be found in DETR Circular 05/99.

Under the Transport Act 2000, local authorities are empowered to designate roads as 'quiet lanes' or 'home zones', thereby placing greater restrictions on the use of them.

Note that since the Road Traffic Act 1991, increasing reliance has been placed on the use of automatic devices for detecting speeding offences and that the procedure of 'conditional offers' of fixed penalties are available for such offences (**see chapter 14**).

The use of reflective or elaborately designed number plates to thwart speed cameras has now been directly addressed in recent regulations (**see chapter 12**).

7.4.1 Speed Limits on Restricted Roads

Section 82 of the Road Traffic Regulation Act 1984 defines a 'restricted road' as:

(1) Subject to the provisions of this section and of section 84(3) of this Act, a road is a restricted road for the purposes of section 81 of this Act if—
   (a) in England and Wales, there is provided on it a system of street lighting furnished by means of lamps placed not more than 200 yards apart;
   (b) (applies to Scotland only).

(2) The traffic authority for a road may direct—
   (a) that the road which is a restricted road for the purposes of section 81 of this Act shall cease to be a restricted road for those purposes, or
   (b) that the road which is not a restricted road for those purposes shall become a restricted road for those purposes.

By virtue of s. 85(5), where a road has such a system of street lamps the lack of any traffic signs specifically saying that the road is not a 'restricted' road will be evidence that it *is* a 'restricted' road.

If there are no such street lamps then there must be traffic signs stating what the speed limit is (s. 85(4)).

7.4.2 Traffic Signs

Traffic signs generally must conform to the Traffic Signs Regulations and General Directions 2002 (SI 2002/3113) (as to which, **see chapter 10**). The specifications relating to signs marking the beginning and end of speed limits on roads are set out in Part II of the Directions (particularly at paras 8–16).

Section 81(1) of the 1984 Act states that it shall be unlawful for a person to drive a motor vehicle on a restricted road at a speed exceeding 30 mph (**see para. 7.4.6**).

Section 84 allows for speed limits to be imposed on roads other than restricted roads and motorways.

7.4.3 Temporary Speed Limits

Section 88 of the 1984 Act provides for both maximum and minimum temporary speed limits to be imposed on certain roads (**see para. 7.4.6**).

Traffic authorities may impose temporary speed *restrictions* in connection with road works or similar operations near to the road which present a danger to the public or serious damage to the highway (see ss. 14–16 of the Road Traffic Regulation Act 1984). These restrictions cannot generally exceed 18 months without approval from the Secretary of State.

As *restrictions* rather than speed limits, offences under this heading do not require notices of intended prosecution (**see chapter 3**) (*Platten* v *Gowing* [1983] Crim LR 184), neither do they require corroboration (**see para. 7.4.6**).

7.4.4 Speed Limits for Particular Classes of Vehicle

Section 86 of the Road Traffic Regulation Act 1984 states:

(1) It shall not be lawful for a person to drive a motor vehicle of any class on a road at a speed greater than the speed specified in Schedule 6 to this Act as the maximum speed in relation to a vehicle of that class.

(2) Subject to subsections (4) and (5) below, the Secretary of State may by regulations vary, subject to such conditions as may be specified in the regulations, the provisions of that Schedule.

(3) Regulations under this section may make different provision as respects the same class of vehicles in different circumstances.

(4) ...

(5) The Secretary of State shall not have power under this section to vary the speed limit imposed by section 81 of this Act.

Schedule 6 of the 1984 Act is set out at **appendix 2**.

7.4.5 Exemption for Police, Fire and Ambulance Purposes

Section 87 of the Road Traffic Regulation Act 1984 states:

> No statutory provisions imposing a speed limit on motor vehicles shall apply to any vehicle on an occasion when it is being used for fire brigade, ambulance or police purposes, if the observance of that provision would be likely to hinder the use of the vehicle for the purpose for which it is being used on that occasion.

KEYNOTE

It is the *purpose* to which the vehicle is being put which matters here. Consequently, a privately-owned vehicle used for police purposes may attract the exemption under s. 87 if it is driven in excess of the speed limit. However, a police vehicle being used for private purposes (i.e. doing some unauthorised 'shopping') would not attract the exemption.

Section 87 only applies to *speed*; it does not absolve drivers of emergency vehicles from any breaches of driving *standards*, see chapter 2.

7.4.6 Proof of Speed

Section 89(2) of the 1984 Act requires that a person prosecuted for driving a motor vehicle at a speed exceeding the limit shall not be convicted solely on the evidence of one witness to the effect that, in his/her opinion, the defendant was exceeding the speed limit. Section 88(7) includes a similar provision for failing to attain a minimum speed limit.

The requirements for corroboration do not apply to general speeding offences on motorways (but they do for special classes of vehicle exceeding motorway speed limits).

Corroboration may be provided by the equipment in a police vehicle or by *Vascar* or similar speed measuring equipment (*Nicholas* v *Penny* [1950] 2 All ER 89). While it may be preferable, it is not *necessary* in such cases to prove the accuracy of the equipment being used (see e.g. *Darby* v *DPP* [1995] RTR 294).

Two police officers may provide sufficient evidence in a case of speeding but the court will decide how much weight (**see Evidence and Procedure, chapter 11**) to give to such evidence. It is important to show that both officers saw the vehicle at exactly the same time (*Brighty* v *Pearson* [1938] 4 All ER 127).

The signal emitted by a hand-held radar speed gun has been deemed *not* to amount to a 'communication' for the purposes of the Wireless Telegraphy Act 1949 and a person intercepting such signals cannot be prosecuted under that legislation (*R* v *Crown Court of Knightsbridge, ex parte Foot* [1999] RTR 21 (**see General Police Duties, chapter 3**). See, however, the offences of obstruction and incitement (**Crime, chapters 3 and 8**).

The Road Traffic Offenders (Prescribed Devices) Order 1999 (SI 1999/162) makes provision for the use of speed cameras that calculate the average speed of a vehicle while passing between two points and allows these readings to be used as proof of the vehicle's speed under s. 20 of the Road Traffic Offenders Act 1988.

The heavy reliance on roadside cameras for enforcing speed limits has given rise to some inventive pleading. While it is possible to delay the whole prosecution process for many months—or even years—by demanding proof of technical and legal data about the cameras, arguing that the driver cannot be ascertained and so forth, such delay is usually simply postponing the inevitable. And for an example of the dim view that the courts will take if, having caused many years delay, a defendant then seeks to rely on that delay as a reason to stay the prosecution against them, see *R (On the Application of Johnson)* v *Stratford Magistrates' Court* [2003] EWHC 353.

7.4.7 Punishment of Speeding Offences

Speeding offences are punishable under the following Acts:

- Contravening speed limits generally (s. 89(1) of the Road Traffic Regulation Act 1984 and sch. 2 to the Road Traffic Offenders Act 1988).
- Contravening motorway speed limits (other than special classes of vehicle) (s. 17(4) of the Road Traffic Regulation Act 1984 and sch. 2 to the Road Traffic Offenders Act 1988).
- Contravening *minimum* speed limits made under s. 88(1)(b) of the Road Traffic Regulation Act 1984 (s. 88(7) of the Road Traffic Regulation Act 1984 and sch. 2 to the Road Traffic Offenders Act 1988).
- Contravening temporary speed *restrictions* made under s. 14 of the Road Traffic Regulation Act 1984 (s. 16(1) of the Road Traffic Regulation Act 1984 and sch. 2 to the Road Traffic Offenders Act 1988).

Section 89(1) of the Road Traffic Regulation Act 1984 states:

> (1) A person who drives a motor vehicle on a road at a speed exceeding a limit imposed by or under any enactment to which this section applies shall be guilty of an offence.

KEYNOTE

The enactments to which this section applies are the Road Traffic Regulation Act 1984 itself *except s. 17(2)* (ordinary classes of vehicles on motorways).

7.5 Motorways

The classes of vehicle which can use motorways at any time are those falling under Class I and Class II in sch. 4 to the Highways Act 1980. These include:

Class I

- Motor cars and motor cycles with an engine/cylinder capacity not less than 50 cc which comply with the Road Vehicles (Construction and Use) Regulations 1986 (**see chapter 9**).
- Heavy and light locomotives and heavy motor cars.
- Trailers drawn by the above.

Class II

- Motor vehicles and trailers authorised to carry abnormal indivisible loads.
- Armed forces vehicles.
- 'Special type' vehicles.

All other vehicles may only use a motorway under the conditions set out in reg. 15 of the Motorways Traffic (England and Wales) Regulations 1982 (SI 1982/1163), as amended.

Regulation 15(1)(b) of the 1982 Regulations prohibits the use of motorways by pedestrians who, if they refuse to leave a motorway, may be arrested for obstructing the highway (**see chapter 8**) (*Reed* v *Wastie* [1972] Crim LR 221).

The Motorways Traffic (England and Wales) Regulations 1982 (SI 1992/1163) set out a number of key definitions including 'hard shoulder', 'carriageway' and 'verge'. They contain further prohibitions on activities such as stopping (reg. 7), reversing (reg. 8) and using the central reservation (reg. 10).

Learner drivers (**see chapter 11**) are excluded from using motorways (by reg. 11), but these will not include drivers who hold ordinary licences but who are learning to drive large goods vehicles and passenger carrying vehicles.

Regulation 12 lists those vehicles which are prohibited from using the outside lane of a motorway (principally goods vehicles exceeding 7.5 tonnes, passenger vehicles over 12 m long and vehicles drawing trailers).

For a list of motorway offences and penalties, **see appendix 3**.

7.5.1 Speeding on Motorways

Schedule 6 to the Road Traffic Regulation Act 1984 sets speed limits for particular classes of vehicles using motorways. Otherwise the offence of exceeding the speed limit on a motorway is punishable under s. 17(4) of the Road Traffic Regulation Act 1984 and sch. 2 to the Road Traffic Offenders Act 1988.

8 Other measures affecting safety

8.1 Introduction

In addition to the legislation regulating driver fitness and competence, there are many other provisions aimed at improving—and enforcing—safety. This chapter addresses some of the more common measures affecting safety, including the parking of vehicles.

8.2 Obstruction

OFFENCE: **Obstruction of the Highway—*Highways Act 1980, s. 137***

• Triable summarily • Fine

(Power of arrest under Police and Criminal Evidence Act 1984, s. 25(3)(d)(v) general conditions)

The Highways Act 1980, s. 137 states:

> (1) If a person, without lawful authority or excuse, in any way wilfully obstructs the free passage along a highway he is guilty of an offence . . .

OFFENCE: **Obstruction on a Road—*Road Vehicles (Construction and Use) Regulations 1986, reg. 103***

• Triable summarily • Fine

(No specific power of arrest)

The Road Vehicles (Construction and Use) Regulations 1986, reg. 103 states:

> No person in charge of a motor vehicle or trailer shall cause or permit the vehicle to stand on a road so as to cause any unnecessary obstruction of the road

OFFENCE: **Obstruction of Street—*Town Police Clauses Act 1847, s. 28***

• Triable summarily • Fourteen days' imprisonment and/or fine

(No specific power of arrest)

The Town Police Clauses Act 1847, s. 28 states:

> Every person who in any street, to the obstruction, annoyance, or danger of the residents or passengers, . . . wilfully interrupts any public crossing, or wilfully causes any obstruction in any public footpath or other public thoroughfare . . .

KEYNOTE

Highway authorities have a statutory duty to maintain highways and to remove obstructions under certain circumstances; this duty can be enforced by individuals serving a statutory notice on the relevant highway authority (see, for example, s. 130 of the Highways Act 1980).

What amounts to an obstruction is a question of fact for the magistrate(s) to decide in each case (*Wade* v *Grange* [1977] RTR 417; see also *DPP* v *Jones* [1999] 2 WLR 625). In arriving at that decision they will have regard to, amongst other things: 'the length of time the obstruction continued, the place where it occurred, the purpose for which it was done and whether it caused an actual as opposed to a potential obstruction' (Lord Parker CJ in *Nagy* v *Weston* [1965] 1 All ER 78).

Although primarily concerned with a trespassory assembly under the Public Order Act 1986 (as to which, **see General Police Duties, chapter 4**), *DPP* v *Jones* (above) raised a number of significant points in relation to obstruction. In deciding that the public's rights in relation to the highway were not restricted to simply passing and re-passing, Lord Irvine said:

> ... the public highway is a place which the public may enjoy for any reasonable purpose, provided the activity in question does not amount to a public or private nuisance and does not obstruct the highway by unreasonably impeding the primary right of the public to pass and re-pass: within these qualifications there is a public right of peaceful assembly on the highway.

His Lordship gave examples of lawful use of the highway as including the handing out of leaflets (see *Hirst & Agu* v *Chief Constable of West Yorkshire* (1985) 85 Cr App R 143), making sketches and collecting for charity.

In addition, their Lordships had been asked to consider the lawful use of the highway in the light of Article 11 of the European Convention on Human Rights, the Article that protects a right of peaceful assembly generally. Lord Irvine stated that to view *any* assembly on the highway as being *prima facie* unlawful (as the Divisional Court had done) would not be compatible with Article 11. This is a useful illustration of how the Convention—and ultimately the Human Rights Act 1998—can be used within any area of existing law. (For a full discussion on the Convention and the 1998 Act, **see General Police Duties, chapter 2**).

Each of the above offences has different elements which are important in selecting the appropriate charge. The Highways Act 1980 offence can be committed by anyone using anything whereas the offence under reg. 103 refers to a person in charge of a motor vehicle or trailer. Similarly, the *place* being obstructed differs from a 'highway', 'road' or 'street' according to the enactment, while the *way* in which that place is obstructed may be either 'wilful' or 'unnecessary'.

Both of the offences, however, appear to require an unreasonable use of the road. Contrast the situation where the power to require vehicles to be moved or the power to move them applies (**see para. 8.6.2**).

Some examples of what might amount to obstruction have included:

- Parking a van in a lay-by to sell food and drink to passers by (*Waltham Forest LBC* v *Mills* [1980] RTR 201).
- Creating a queue of vehicles at a security gate at the entry to a factory (*Lewis* v *Dickson* [1976] RTR 431).
- Putting displays out on the footpath in front of a shop (*Hertfordshire County Council* v *Bolden* (1987) 151 JP 252).
- Street entertainers in a shopping street (*Waite* v *Taylor* (1985) 149 JP 551).

Where a person is convicted of an offence of obstruction under s. 137 of the 1980 Act above and the obstruction is not removed, the court can make an order requiring the person to remove the cause of the obstruction, provided that it is in that person's power to do so (s. 137ZA). Such an order can be made in addition to, or in place of any other punishment and will specify a reasonable period within which the cause of the obstruction must be removed. If the person fails without reasonable excuse to comply with such an order, he/she commits a summary offence punishable by a fine which will increase incrementally for every day that that the obstruction remains.

A refusal to move a vehicle that is obstructing the highway when requested by a police officer can amount to an 'obstruction' of the officer under what is now s. 89(2) of the Police Act 1996 (**see Crime, chapter 8**) (*Gelberg* v *Miller* [1961] 1 WLR 153).

A continuing or repeated obstruction may amount to the offence of a public nuisance (as to which, **see General Police Duties, chapter 3**).

8.2.1 Terrorism

Additional powers to regulate the parking of vehicles by the police are found in the Terrorism Act 2000. Where it appears to any officer of or above the rank of assistant chief constable/commander (in relation to provincial forces and the Metropolitan/City of London police respectively) that it is expedient to do so, in order to prevent acts of terrorism, he/she may authorise the use of stop and search powers in a locality for a period not exceeding 28 days (ss. 44 and 46). Similar powers are given to officers of the specified ranks to impose prohibitions or restrictions on parking in a specified area (s. 48). These prohibitions and restrictions have to be accompanied by the placing of traffic signs on the relevant road (s. 49). Section 49 goes on to give the police powers to suspend parking places and extends the power under s. 99 of the Road Traffic Regulation Act 1984 (removal of illegally parked vehicles; **see para. 8.6.2**) to cover such situations.

Anyone parking any vehicle in contravention of a prohibition or restriction imposed under s. 48 commits a summary offence punishable by a fine (s. 51(1)). Although it is a defence for a person charged with an offence under this section to prove that he/she had a reasonable excuse for the act or omission, possession of a current disabled person's badge (**see para. 8.4.4**) is not itself a 'reasonable excuse' for these purposes (s. 51(2) and (3)).

If the driver or person in charge of any vehicle:

- permits it to remain at rest in contravention of any such prohibition or restriction, or
- fails to move the vehicle when ordered to do so by a constable *in uniform*

he/she commits a further summary offence punishable with a fine and/or three months' imprisonment (s. 51(3)).

For a discussion of terrorism generally, **see General Police Duties, chapter 4**.

8.2.2 Stopping up highways to prevent crime

The Highways Act 1980 has been amended to allow local highway authorities to stop up or divert highways where it is expedient to do so for the purposes of preventing or reducing crime which would otherwise disrupt the life of the community. The amendments (made by the Countryside and Rights of Way Act 2000) empower highway authorities to make a special order, either stopping up or diverting the relevant highway, where they are satisfied that premises adjoining or adjacent to the highway are affected by high levels of crime and that the existence of the highway is facilitating the persistent commission of criminal offences. If the relevant highway crosses school land an order can be made if it is expedient for the purposes of protecting the pupils or staff from violence or threats of violence, or from harassment, alarm or distress arising from unlawful activity. Neither a stopping up order (under s. 118B of the 1980 Act), nor a diversion order (under s. 119B) can be made until the highway authority has consulted the police authority for the area in which the highway lies. These orders must be submitted for confirmation by the Secretary of State. The majority of these powers came into effect in February 2003. For full details, see sch. 6 to the Countryside and Rights of Way Act 2000.

8.3 Causing Danger

OFFENCE: **Causing Danger to Other Road Users—*Road Traffic Act 1988, s. 22A***

- Triable either way • Seven years' imprisonment and/or a fine on indictment; six months' imprisonment and/or a fine summarily

(Arrestable offence)

The Road Traffic Act 1988, s. 22A states:

(1) A person is guilty of an offence if he intentionally and without lawful authority or reasonable cause—
(a) causes anything to be on or over a road, or
(b) interferes with a motor vehicle, trailer or cycle, or
(c) interferes (directly or indirectly) with traffic equipment,
in such circumstances that it would be obvious to a reasonable person that to do so would be dangerous.

(2) In subsection (1) above 'dangerous' refers to danger either of injury to any person while on or near a road, or of serious damage to property on or near a road; and in determining for the purposes of that subsection what would be obvious to a reasonable person in a particular case, regard shall be had not only to the circumstances of which he could be expected to be aware but also to any circumstances shown to have been within the knowledge of the accused.

(3) In subsection (1) above 'traffic equipment' means—
(a) anything lawfully placed on or near a road by a highway authority;
(b) a traffic sign lawfully placed on or near a road by a person other than a highway authority;
(c) any fence, barrier or light lawfully placed on or near a road—
(i) in pursuance of section 174 of the Highways Act 1980, or section 65 of the New Roads and Street Works Act 1991 (which provide for guarding, lighting and signing in streets where works are undertaken), or
(ii) by a constable or a person acting under the instructions (whether general or specific) of a chief officer of police.

(4) For the purposes of subsection (3) above anything placed on or near a road shall unless the contrary is proved be deemed to have been lawfully placed there.

KEYNOTE

This offence requires a defendant to act both intentionally and without lawful authority. Although crimes requiring 'intention' are generally more difficult to prove than those capable of commission by some lesser state of mind (such as recklessness) (**see Crime, chapter 1**), the intention in s. 22A only applies to the causing or interfering described. There is no need to show that the defendant *intended* to create danger—only that it would have been obvious to a reasonable person that to do so would be dangerous.

Like the requirements for dangerous driving under ss. 1 and 2 of the 1988 Act, the test here is primarily *objective* (the 'reasonable person' test) but it also has a *subjective* element in that regard will be had to any circumstances shown to be known to the defendant.

A 'road' for the purpose of this offence does not include a footpath (or bridleway) (s. 22A(5)).

8.3.1 Dangerous Activities on Highways

OFFENCE: **Dangerous Activity on Highways—*Highways Act 1980, ss. 161 to 162***

- Triable summarily • Fine

(No specific power of arrest)

The Highways Act 1980, ss. 161 to 162 state:

161.—(1) If a person, without lawful authority or excuse, deposits any thing whatsoever on a highway in consequence of which a user of the highway is injured or endangered, that person is guilty of an offence...

(2) If a person, without lawful authority or excuse—
(a) lights any fire on or over a highway which consists of or comprises a carriageway; or
(b) discharges any firearm or firework within [15.24 metres] 50 feet of the centre of such a highway,
and in consequence a user of the highway is injured, interrupted or endangered, that person is guilty of an offence...

(3) If a person plays at football or any other game on a highway to the annoyance of a user of the highway he is guilty of an offence...

(4) If a person, without lawful authority or excuse, allows any filth, dirt, lime or other offensive matter or thing to run or flow on to a highway from any adjoining premises, he is guilty of an offence...

161A.—(1) If a person—
(a) lights a fire on any land not forming part of a highway which consists of or comprises a carriageway; or
(b) directs or permits a fire to be lit on any such land,
and in consequence a user of any highway which consists of or comprises a carriageway is injured, interrupted or endangered by, or by smoke from, that fire or any other fire caused by that fire,
that person is guilty of an offence...

162.— A person who for any purpose places any rope, wire or other apparatus across a highway in such a manner as to be likely to cause danger to persons using the highway is, unless he proves that he had taken all necessary means to give adequate warning of the danger, guilty of an offence...

KEYNOTE

For the definition of 'highway', **see chapter 1**.

For the offence of throwing fire or fireworks into a street or highway under s. 80 of the Explosives Act 1875), **see General Police Duties, chapter 3**.

'Carriageway' means a way constituting or comprised in a highway, being a way (other than a cycle track) over which the public have a right of way for the passage of vehicles (s. 329 of the 1980 Act).

The offences at ss. 161(1), (2) and (3) and s. 161A(1) are all offences of 'consequence', that is, you must show the relevant consequence (e.g. injury, annoyance, etc.).

Although no specific power of arrest exists, the general arrest condition under s. 25(3)(d)(v) of the Police and Criminal Evidence Act 1984 may apply (**see General Police Duties, chapter 2**).

For more serious offences involving a threat to the safety of motorists, see above. For other offences involving fireworks and causing nuisance generally, **see General Police Duties, chapter 3**.

8.4 Parking

8.4.1 Parking on Verges

OFFENCE: **Parking of Heavy Commercial Vehicles on Verges—*Road Traffic Act 1988, s. 19***

- Triable summarily • Fine

(No specific power of arrest)

The Road Traffic Act 1988, s. 19 states:

(1) Subject to subsection (2) below, a person who parks a heavy commercial vehicle (as defined in section 20 of this Act) wholly or partly—
(a) on the verge of a road, or
(b) on any land situated between two carriageways and which is not a footway, or
(c) on a footway,
is guilty of an offence.

Defence

The Road Traffic Act 1988, s. 19 states:

(2) A person shall not be convicted of an offence under this section in respect of a vehicle if he proves to the satisfaction of the court—
(a) that it was parked in accordance with permission given by a constable in uniform, or

(b) that it was parked in contravention of this section for the purpose of saving life or extinguishing fire or meeting any other like emergency, or

(c) that it was parked in contravention of this section but the conditions specified in subsection (3) below were satisfied.

(3) The conditions mentioned in subsection (2)(c) above are—

(a) that the vehicle was parked on the verge of a road or on a footway for the purpose of loading or unloading, and

(b) that the loading or unloading of the vehicle could not have been satisfactorily performed if it had not been parked on the footway or verge, and

(c) that the vehicle was not left unattended at any time while it was so parked.

(4) In this section 'carriageway' and 'footway', in relation to England and Wales, have the same meanings as in the Highways Act 1980.

KEYNOTE

'Heavy commercial vehicle' means any goods vehicle which has an operating weight exceeding 7.5 tonnes (Road Traffic Act 1988, s. 20(1)).

The operating weight of a goods vehicle for the purposes of this section is:

- in the case of a motor vehicle not drawing a trailer or in the case of a trailer, its maximum laden weight,
- in the case of an articulated vehicle, its maximum laden weight (if it has one) and otherwise the aggregate maximum laden weight of all the individual vehicles forming part of that articulated vehicle, and
- in the case of a motor vehicle (other than an articulated vehicle) drawing one or more trailers, the aggregate maximum laden weight of the motor vehicle and the trailer or trailers attached to it.

For the purposes of this section: a 'carriageway' is a part of the highway (other than a cycle track) over which members of the public have a right of way for vehicles; a 'footway' is part of a highway that also comprises a carriageway, being a way over which the public has a right to pass on foot only (simply the pavement in most cases) (s. 329(1) of the Highways Act 1980).

8.4.2 Leaving Vehicles in Dangerous Positions

OFFENCE: **Leaving Vehicles in Dangerous Positions—*Road Traffic Act 1988, s. 22***

- Triable summarily • Fine • Discretionary disqualification

(No specific power of arrest)

The Road Traffic Act 1988, s. 22 states:

If a person in charge of a vehicle causes or permits the vehicle or a trailer drawn by it to remain at rest on a road in such a position or in such condition or in such circumstances as to involve a danger of injury to other persons using the road, he is guilty of an offence.

KEYNOTE

This offence involves presenting a danger of injury to other road users by the *position*, *condition* or *circumstances* of the vehicle/trailer.

The danger presented by the condition or circumstances of the vehicle is not confined to occasions when it is stationary but will also apply to a vehicle/trailer which presents a danger by moving (such as where a driver fails to set the handbrake (*Maguire* v *Crouch* (1940) 104 JP 445)).

The risk must be to 'other persons using the road'.

8.4.3 Parking Regulations

Much of the responsibility for regulating the parking of vehicles on roads falls on local authorities. The Road Traffic Regulation Act 1984, ss. 45–56, makes provision for the parking of motor vehicles on highways and for the operation of parking meters. The designation of parking places by local authorities is provided for under s. 46 while s. 47 creates an offence of failing to comply with the conditions of such a designated parking place.

Special provision is made in relation to parking in London under part IV of the Greater London Authority Act 1999.

The Road Traffic Regulation Act 1984, ss. 1–8 allows for the creation of 'no waiting' orders. When creating these orders, local authorities must follow the procedure laid out in the Local Authorities' Traffic Orders (Procedure) (England and Wales) Regulations 1989 (SI 1989/1120).

It is worth noting that a council has a statutory duty (under s. 21 of the Housing Act 1985) to manage, regulate and control its houses. This has been held to empower a council to enforce regulations for the control of parking—for instance by clamping vehicles where appropriate (see *Akumah* v *Hackney London Borough Council* [2002] EWCA Civ 582). A council also has subsidiary powers to do anything that is conducive or incidental to the discharge of its functions (see s. 111 of the Local Government Act 1972). As a result, a council has the power to enforce measures for the proper regulation of parking without the need to make specific by-laws first.

OFFENCE: **Contravention of Traffic Regulation Order—*Road Traffic Regulation Act 1984, s. 5(1) (outside London); s. 8(1) (inside London)***

- Triable summarily • Fine

(No specific power of arrest)

The Road Traffic Regulation Act 1984, ss. 5 and 8 state:

> 5.—(1) A person who contravenes a traffic regulation order, or who uses a vehicle, or causes or permits a vehicle to be used in contravention of a traffic regulation order, shall be guilty of an offence.
>
> 8.—(1) Any person who acts in contravention of, or fails to comply with, an order under section 6 of this Act shall be guilty of an offence.
>
> (1A) Subsection (1) above does not apply in relation to any order under section 6 of this Act so far as it designates any parking places.

KEYNOTE

In order to be effective, signs showing 'no waiting' areas must be properly marked and displayed as required by the Road Traffic Regulation Act 1984, s. 64 and the Traffic Signs Regulations 2002 (SI 2002/3113). If they are not then no offence of contravention can be prosecuted (see *Hassan* v *DPP* [1992] RTR 209).

8.4.4 Orange Badge Scheme: Disabled Drivers

Local authorities are empowered to issue 'orange badges' under the Chronically Sick and Disabled Persons Act 1970 (as amended) for motor vehicles driven by, or used for the carriage of disabled persons.

Exemptions for disabled people driving or being carried in vehicles using an orange badge are covered by the Local Authorities' Traffic Orders (Exemptions for Disabled Persons) (England and Wales) Regulations 1986 (SI 1986/178), as amended. These Regulations (which do not apply in certain areas within London) require exemptions in certain local no waiting and parking orders. Their effect is generally to increase or remove altogether the waiting time allowed for badge holders, and to allow such people to park without charge in certain areas. The scheme applies only to orders made under the Road Traffic Regulation

Act 1984 and does not provide some form of general exemption from other legislation controlling traffic.

The regulation of the so-called 'orange badge' scheme *in England* falls under the Disabled Persons (Badges for Motor Vehicles) (England) Regulations 2000 (SI 2000/682). These Regulations came into force on 1 April 2000. In relation to Wales, the relevant law can be found in the Disabled Persons (Badges for Motor Vehicles) (Wales) Regulations 2000 (SI 2000/1786 (W. 123)) which came into force on 1 July 2000.

The 2000 Regulations were introduced to give effect to an EC Council Recommendation for a Community-model parking card for people with disabilities. These badges will have a blue background. While existing orange badges will remain in force until they are due for renewal, all new applicants for badges made under the Chronically Sick and Disabled Persons Act 1970 will be issued with the blue badge.

In addition to continuing the individual badge scheme, the 2000 Regulations introduce a new 'institutional' badge issued to a relevant institution for its motor vehicles when they are being used to carry people who are disabled within the meaning of the Regulations. The 2000 Regulations go on to make provision for the issuing, refusal, withdrawal and display of the individual and institutional badges.

For the general effects of the Disability Discrimination Act 1995, **see General Police Duties, chapter 12**.

OFFENCE: **Wrongful Use of Disabled Person's Badge—*Road Traffic Regulation Act 1984, s. 117***

Triable summarily • Fine

(No specific power of arrest)

The Road Traffic Regulation Act 1984, s. 117 states:

(1) A person who is guilty of an offence in relation to a motor vehicle under a provision of this Act other than this section ("the first offence") is also guilty of an offence under this section if the conditions specified in subsection (2) below are satisfied.

(2) The conditions mentioned in subsection (1) above are that at the time of the commission of the first offence—

(a) a disabled person's badge was displayed on the motor vehicle;

(b) he was using the motor vehicle in circumstances where a disabled person's concession would be available to a disabled person's vehicle; and

(c) the vehicle was not being used either by the person to whom the badge was issued or under section 21(4) (institutional use) of the Chronically Sick and Disabled Persons Act 1970.

(3) In this section—

'disabled person's concession' means—

(a) an exemption from an order under this Act given by reference to disabled persons' vehicles; or

(b) a provision made in any order under this Act for the use of a parking place by disabled persons' vehicles.

KEYNOTE

A local authority may require the return of a badge and may refuse to issue such a badge if the person concerned has held and subsequently misused it in a way which has led to at least three relevant convictions.

A driver who commits a parking offence and misuses an orange badge at the same time commits both the relevant parking offence *and* an offence under s. 117 above.

8.4.5 Removal and Immobilisation of Parked Vehicles

Sections 104–106 of the Road Traffic Regulation Act 1984 provide for the clamping of vehicles on roads using an 'approved' device (listed at s. 104(12A)). The only roads currently affected by this power are in central London although the Secretary of State may identify further areas in which it will apply.

Wheel clamping on private land is a different matter. If an occupier of land decides to enforce a civil remedy of detaining the property of trespassers until compensation is paid (i.e. by clamping their vehicle), that is a civil matter. Where an occupier of private land displayed a clear notice that vehicles parking without authority would be clamped, the Divisional Court held that anyone who subsequently parked there without authority consented to their car being clamped. The vehicle owner could not be excused for causing criminal damage (**see Crime, chapter 14**) to the clamp by trying to remove it (*Lloyd* v *DPP* [1991] Crim LR 904).

The Private Security Industry Act 2001, which received Royal Assent on 11 May 2001, creates a Security Industry Authority whose remit will include the regulation of wheel clamping on private land and the introduction of a code of practice for clamping firms. The framework allowing for regulations to be made came into force in April 2003.

Note that under the Vehicle Excise Duty (Immobilisation, Removal and Disposal of Vehicles) Regulations 1997 (SI 1997/2439) vehicles may be clamped if they are not taxed (**see chapter 12**).

Interestingly, in Scotland wheel clamping has been held to be 'extortion' (*Black* v *Carmichael* [1992] SLT 897) and there are strong arguments to suggest that the clamping of vehicles without statutory authority is unlawful (see the article by Kruse J., *Legal Action*, July 1998, p. 27).

Special provisions in relation to the disposal and immobilisation of vehicles in London are set out at s. 101 of the Road Traffic Regulation Act 1984. Responsibility for regulating traffic and parking in London has been transferred to 'Traffic for London' (as to which, **see chapter 10**) by the Greater London Authority Act 1999.

8.5 Tampering with and Getting on to Vehicles

OFFENCE: **Tampering with Motor Vehicles—*Road Traffic Act 1988, s. 25***

- Triable summarily • Fine

(No specific power of arrest)

The Road Traffic Act 1988, s. 25 states:

> If, while a motor vehicle is on a road or on a parking place provided by a local authority, a person—
> (a) gets on to the vehicle, or
> (b) tampers with the brake or other part of its mechanism,
> without lawful authority or reasonable cause he is guilty of an offence

KEYNOTE

This offence, which only applies to 'motor vehicles' and not trailers, can only be committed where the vehicle is on a road or local authority parking place.

What constitutes a part of its mechanism—other than the brake—is open to interpretation but it would appear that anything falling within the ordinary meaning of 'mechanism' would suffice.

If a defendant has got on to or tampered with the vehicle in order to steal it or part of its load then the offence under s. 9 of the Criminal Attempts Act 1981 (interfering) may be appropriate (**see Crime, chapter 3**).

This is a specified offence for the purposes of s. 3(4)(b) of the Vehicles (Crime) Act 2001.

If a defendant gets on to a *moving* vehicle—or trailer—he/she may commit the following offence.

OFFENCE: **Holding or Getting on to Vehicle in Motion—*Road Traffic Act 1988, s. 26***
- Triable summarily • Fine

(No specific power of arrest)

The Road Traffic Act 1988, s. 26 states:

(1) If, for the purpose of being carried, a person without lawful authority or reasonable cause takes or retains hold of, or gets on to a motor vehicle or trailer while in motion on a road he is guilty of an offence.

(2) If, for the purpose of being drawn, a person takes or retains hold of a motor vehicle or trailer while in motion on a road he is guilty of an offence.

KEYNOTE

For this offence the vehicle or trailer must both be in motion and on a road.

8.6 Abandoning Vehicles

OFFENCE: **Abandoning Motor Vehicle—*Refuse Disposal (Amenity) Act 1978, s. 2(1)***
- Triable summarily • Fine

(No specific power of arrest)

The Refuse Disposal (Amenity) Act 1978, s. 2 states:

(2) Any person who, without lawful authority,—
   (a) abandons on any land in the open air, or on any other land forming part of a highway, a motor vehicle or anything which formed part of a motor vehicle and was removed from it in the course of dismantling the vehicle on the land; or
   (b) abandons on any such land any thing other than a motor vehicle, being a thing which he has brought to the land for the purpose of abandoning it there,

shall be guilty of an offence.

KEYNOTE

For police powers in relation to bringing vehicles on to land under the Criminal Justice and Public Order Act 1994, see General Police Duties, chapter 9.

8.6.1 Duty of Local Authority to Remove Abandoned Vehicles

Section 3 of the Refuse Disposal (Amenity) Act 1978 states:

(1) Where it appears to a local authority that a motor vehicle in their area is abandoned without lawful authority on any land in the open air or on any other land forming part of a highway, it shall be the duty of the authority, subject to the following provisions of this section, to remove the vehicle.

'Motor vehicle' is defined under s. 11 of the Act as:

. . . a mechanically propelled vehicle intended or adapted for use on roads, whether or not it is in a fit state for such use, and includes any trailer intended or adapted for use as an attachment to such a vehicle, any chassis or body, with or without wheels, appearing to have formed part of such a vehicle or trailer, and anything attached to such a vehicle or trailer.

KEYNOTE

Before it can remove a vehicle under this section, a local authority must follow the requirements of the Act and of the Removal and Disposal of Vehicles Regulations (SI 1986 No. 183) in relation to affixing of notices advising the owner (of both the vehicle, and the land if occupied) of its intention. In order to combat the growing problem of abandoned cars, the notice period required to be given by a local authority in England in relation to a vehicle which appears to have been abandoned and ought to be destroyed has been reduced from seven days to 24 hours (see reg. 10).

Where an authority has removed a vehicle which is not in such a condition that it ought to be destroyed, and the authority has located the owner, the period during which the owner must remove the vehicle after service of the relevant notice is now seven days (see reg. 14).

8.6.2 Police Power to Remove Vehicles from Roads

As discussed above (**see para. 8.2**), the police have particular powers to regulate the parking or leaving of vehicles under certain circumstances. Similarly, under certain circumstances police officers can require a vehicle to be removed from a particular road or place.

Regulation 3 of the Removal and Disposal of Vehicles Regulations 1986 above states:

> (2) A constable may require the owner, driver or other person in control or in charge of any vehicle to which this Regulation applies to move or cause to be moved the vehicle and any such requirement may include a requirement that the vehicle shall be moved from that road to a place which is not on that or any other road, or that the vehicle shall not be moved to any such road or to any such position on a road as may be specified.

KEYNOTE

These regulations are made under s. 99 of the Road Traffic Regulation Act 1984.

The vehicles to which this regulation applies are vehicles which:

- have broken down, or been permitted to remain at rest, on a road in such a position or in such condition or in such circumstances as to cause obstruction to persons using the road or as to be likely to cause danger to such persons, or
- have been permitted to remain at rest or have broken down and remained at rest on a road in contravention of a prohibition or restriction contained in, or having effect under, any of the enactments mentioned in sch. 1 to these Regulations.

Note that this Regulation is not restricted to 'motor' vehicles and it applies to any vehicle, whatever its condition, including chassis or bodies, with or without wheels appearing to have formed part of a vehicle (Road Traffic Regulation Act 1984, s. 99(5)).

'Obstruction' for these purposes is not to be construed in the same way as an obstruction under the Highways Act 1980 or reg. 103 of the Road Vehicles (Construction and Use) Regulations 1986 (**see para. 8.2**). There, an obstruction requires an unreasonable use of the road or highway. Under reg. 3 above the requirement to move the vehicle can be made, not only where the vehicle is *actually* obstructing other traffic, but also where it is *potentially* obstructing other road users that may be expected at some time in the future (*Carey* v *Chief Constable of Avon and Somerset* [1995] RTR 405). This power can be used if the offending vehicle is considered (by the officer) to be 'obstructing the passage of road users along the highway; hindering them or preventing them from getting past' (per Hutchinson LJ in *Carey* above).

Regulation 4A provides traffic wardens with a power to remove or arrange for the removal of vehicles to which reg. 3 applies under certain circumstances. The powers in regulations made under s. 99 of the Road Traffic Regulation Act 1984 are among those that can be conferred on a Community Support Officer designated under sch. 4 to the Police Reform Act 2002 and a person accredited under sch. 5 to that Act (**see General Police Duties, chapter 2**).

Schedule 1 lists most offences relating to unlawful parking (e.g. 'no waiting' areas, pedestrian crossing areas, etc.).

These provisions extend to vehicles parked in contravention of an order made under s. 48 of the Terrorism Act 2000 (see para. 8.2.1).

Regulation 4 states:

Where a vehicle—

(a) is a vehicle to which Regulation 3 of these Regulations applies, or

(b) having broken down on a road or on any land in the open air, appears to a constable to have been abandoned without lawful authority, or

(c) has been permitted to remain at rest on a road or on any land in the open air in such a position or in such condition or in such circumstances as to appear to a constable to have been abandoned without lawful authority,

then, subject to the provisions of sections 99 and 100 of the Road Traffic Regulation Act 1984, a constable may remove or arrange for the removal of the vehicle, and, in the case of a vehicle which is on a road he may remove it or arrange for its removal from that road to a place which is not on that or any other road, or may move it or arrange for its removal to another position on that or another road.

KEYNOTE

Section 99 of the Road Traffic Regulation Act 1984 provides for the removal of vehicles illegally parked or abandoned, while s. 100 deals with the interim disposal of such vehicles.

This power might be used where an officer wishes to move a vehicle and either cannot find the driver or the driver refuses to move it.

Regulation 4 also empowers a police officer to remove vehicles that have broken down on land in the open air.

Regulation 4 confers two powers on the police officer; a personal power to remove the vehicle and a power to arrange for the vehicle's removal by another (*Rivers* v *Cutting* [1983] RTR 105). If the officer arranges for a reputable contractor to remove the vehicle under this power and chooses that contractor with reasonable care, the officer is not liable for any damage caused by the contractor.

In a case involving the theft of a car which had been abandoned by the thieves, the Court of Appeal reviewed the use of these powers by police officers. The Court held that the correct approach under s. 99 of the Act and reg. 4 of the Regulations was not whether a vehicle had been abandoned, but whether, in the wording of reg. 4, the vehicle had been *left in a position so as to appear to the police officer* that it had been abandoned (*Clarke* v *Chief Constable of the West Midlands Police* [2002] RTR 5).

The powers conferred by regulations made under s. 99 of the Road Traffic Regulation Act 1984 are among those that can be conferred on a designated person under sch. 4 to the Police Reform Act 2002 (**see General Police Duties, chapter 2**).

8.7 Off-road Driving

OFFENCE: **Driving Motor Vehicle on Land other than a Road—*Road Traffic Act 1988, s. 34(1)***

- Triable summarily • Fine

(No specific power of arrest)

The Road Traffic Act 1988, s. 34 states:

(1) Subject to the provisions of this section, if without lawful authority a person drives a mechanically propelled vehicle—

(a) on to or upon any common land, moorland or land of any other description, not being land forming part of a road, or

(b) on any road being a footpath, bridleway or restricted byway,

he is guilty of an offence.

KEYNOTE

Where a constable in uniform has reasonable grounds for believing that a mechanically propelled vehicle is being used or has been used on any occasion in a manner which contravenes s. 34, and is causing (or is likely to cause), alarm, distress or annoyance to members of the public, he/she has the powers set out in s. 59 of the Police Reform Act 2002 (as to which, **see chapter 2**).

8.7.1 Defence

Section 34 goes on to state:

> (3) It is not an offence under this section to drive a mechanically propelled vehicle on any land within fifteen yards of a road, being a road on which a motor vehicle may lawfully be driven, for the purpose only of parking the vehicle on that land.

KEYNOTE

This offence has been reworded by the Countryside and Rights of Way Act 2000. The most significant change to the offence is that it has been extended to include 'mechanically propelled vehicles' rather than 'motor vehicles' (for these definitions, **see chapter 1**). This extension means that a broader class of vehicles will now be covered by the offence. The other addition to this offence is that of a 'restricted byway', a new definition that broadly means a highway over which the public have rights of way on foot, on horseback or using vehicles other than mechanically propelled vehicles.

This exception only applies if the driver's purpose in driving on the land is to park on it. Although s. 34(3) could be interpreted as limiting such driving to 15 yards the point is unclear. If, however, the driver's purpose is to do something other than park on the land (e.g. to turn round) the offence will be made out. The section includes any land 'of any other description' which is not part of a road.

There is a further, general defence under s. 34 which states:

> (4) A person shall not be convicted of an offence under this section with respect to a vehicle if he proves to the satisfaction of the court that it was driven in contravention of this section for the purpose of saving life or extinguishing fire or meeting any other like emergency.

KEYNOTE

Note that it is still a summary offence under the Highways Act 1835, s. 72 wilfully to ride or drive on the footway and the offence is not limited to 'motor vehicles'. 'Wilfully' means doing so on purpose and any accidental or inadvertent driving onto the footway would not come under this offence (*Fearnley* v *Ormsby* (1879) 43 JP 384).

8.8 Skips

OFFENCE: **Depositing Builders' Skip on the Highway—*Highways Act 1980, s. 139***

- Triable summarily • Fine

(No specific power of arrest)

The Highways Act 1980, s. 139 states:

> (1) A builder's skip shall not be deposited on a highway without the permission of the highway authority for the highway.

(2) A permission under this section shall be a permission for a person to whom it is granted to deposit, or cause to be deposited, a skip on the highway specified in the permission, and a highway authority may grant such permission either unconditionally or subject to such conditions as may be specified in the permission including, in particular, conditions relating to—
   (a) the siting of the skip;
   (b) its dimensions;
   (c) the manner in which it is to be coated with paint and other material for the purpose of making it immediately visible to oncoming traffic;
   (d) the care and disposal of its contents;
   (e) the manner in which it is to be lighted or guarded;
   (f) its removal at the end of the period of permission.

(3) ...

(4) Where a builder's skip has been deposited on a highway in accordance with a permission granted under this section, the owner of the skip shall secure—
   (a) that the skip is properly lighted during the hours of darkness and, where regulations made by the Secretary of State under this section require it to be marked in accordance with the regulations (whether with reflecting or fluorescent material or otherwise), that it is so marked;
   (b) that the skip is clearly and indelibly marked with the owner's name and with his telephone number or address;
   (c) that the skip is removed as soon as practicable after it has been filled;
   (d) that each of the conditions subject to which that permission was granted is complied with;
   and, if he fails to do so, he is, subject to subsection (6) below, guilty of an offence...

KEYNOTE

Any permission has to be given in writing and it is an offence to fail to comply with any condition imposed upon that permission. 'Blanket' permissions are not permitted (*York District Council* v *Poller* [1976] RTR 37) and the offence will be complete if the owner fails to remove the skip after the permission has expired (*Craddock* v *Green* [1983] RTR 479).

Under s. 139(11) skips for this purpose are containers:

> ...designed to be carried on a road vehicle and to be placed on a highway or other land for the storage of builders' materials, or for the removal and disposal of builders' rubble, waste, household and other rubbish or earth...

KEYNOTE

They must also comply with the Builders' Skips (Markings) Regulations 1984 (SI 1984/1933).

The owner of a skip that is the subject of a hiring or hire purchase agreement is defined under s. 139(11) as the person in possession of it, provided that the agreement is for a period of *not less than* one month.

8.8.1 Defence

Section 139 of the 1980 Act states:

(6) In any proceedings for an offence under this section it is a defence, subject to subsection (7) below, for the person charged to prove that the commission of the offence was due to the act or default of another person and that he took all reasonable precautions and exercised all due diligence to avoid the commission of such an offence by himself or any person under his control.

8.8.2 Police Powers

Under s. 140 of the 1980 Act a constable in uniform may require the removal of a skip from the highway. Failure to do so is an offence under s. 140(3) but the requirement must be made *in person*. If the requirement to remove the skip is made by the officer telephoning the owner, the offence will not be made out (*R* v *Worthing Justices, ex parte Waste Management Ltd* [1988] Crim LR 458).

8.9 Pedestrian Crossings

Section 25 of the Road Traffic Regulation Act 1984 allows the Secretary of State to make regulations with respect to pedestrian crossings.

The current regulations are the Zebra, Pelican and Puffin Pedestrian Crossings Regulations and General Directions 1997 (SI 1997/2400) and these regulations are the ones referred to throughout the following section. Combined pedestrian and cyclist crossings (Toucan crossings) come under reg. 49 of the Traffic Signs Regulations 2002 (SI 2002/3113) (**see chapter 10**).

OFFENCE: **Contravening Regulations made under s. 25 of the Road Traffic Regulation Act 1984—*Road Traffic Regulation Act 1984, s. 25(5)***

- Triable summarily • Fine • Discretionary disqualification

(No specific power of arrest)

The Road Traffic Regulation Act 1984, s. 25 states:

(5) A person who contravenes any regulations made under this section shall be guilty of an offence.

KEYNOTE

Offences committed against the following crossings regulations will generally be prosecuted under the Road Traffic Regulation Act 1984, s. 25. However, some behaviour will involve a breach of a regulation made under *s. 64* of the Road Traffic Regulation Act 1984 (e.g. ignoring a crossing light). In such cases the offence will be triable summarily, attracting a fine but no endorsement or liability to be disqualified (see sch. 2 to the Road Traffic Offenders Act 1988, **appendix 3**).

8.9.1 The Regulations

Somewhat tortuously, the puffin crossing gets its name from the new Pedestrian User-friendly Intelligent crossing! Puffin crossings differ from their less intelligent counterparts by incorporating a pedestrian sensor that delays traffic as long as is necessary while people are on the crossing but not once the crossing is clear.

The Zebra, Pelican and Puffin Pedestrian Crossings Regulations and General Directions 1997 (SI 1997/2400) set out the relevant requirements for each type of crossing, together with the obligations on drivers when at a crossing.

Regulation 3 provides the many definitions found within the 1997 Regulations. The various dimensions of each type of crossing and their respective markings are laid out in regs 5–7 and in schs 1–4.

Under regs 8 and 9, provision is made for traffic authorities to add further apparatus or equipment at crossings (e.g. equipment to help people with disabilities).

Although the prescribed dimensions and markings must be adopted by traffic authorities when erecting such crossings, minor irregularities which do not materially affect the appearance or proper operation of a crossing will not prevent it from being treated as 'a crossing' for the purposes of offences and traffic regulation (reg. 10).

8.9.2 Lights at Crossings

Lights placed at crossings are 'traffic signs' (**see chapter 10**) for the purposes of the Road Traffic Regulation Act 1984 (reg. 11).

Regulations 12 and 13 set out the significance of the lights and signals at pelican and puffin crossings respectively. These regulations state what meanings each signal shall have and what action may or may not be taken by drivers on seeing those signals.

Special provision is made within regs 12 and 13 in relation to vehicles being used for the purposes of:

- police;
- fire brigade;
- ambulance; or
- national blood service.

If a vehicle is being used for one of those purposes *and* if the observance of the steady amber or red signal (or the red-with-amber at a puffin crossing) would be likely to hinder the use of the vehicle for that purpose, the driver of the vehicle may proceed beyond the 'stop' line under certain conditions (regs 12(e) and 13(f)).

Those conditions are that the driver:

- accords precedence to any pedestrian who is on the carriageway within the limits of the crossing; *and*
- does not proceed in a manner *or* at a time likely to endanger any person/vehicle approaching or waiting at the crossing; or
- does not proceed in a manner *or* at a time likely to cause the driver of any such vehicle to change its speed or course in order to avoid an accident.

For a discussion of the standards of police driving in general, **see para. 2.10**.

8.9.3 Give-way Lines at Zebra Crossings

Regulation 14 of the 1997 Regulations states:

> A give-way line included in the markings placed pursuant to regulation 5(1)(b) and Part II of Schedule 1 shall convey to vehicular traffic proceeding towards a Zebra crossing the position at or before which a vehicle should be stopped for the purpose of complying with regulation 25 (precedence of pedestrians over vehicles at Zebra crossings).

8.9.4 Pedestrian Light Signals

Regulation 15 of the 1997 Regulations states:

> (1) The significance of the red and steady green pedestrian light signals whilst they are illuminated at a Pelican crossing and of the red and green figures on a pedestrian demand unit whilst they are illuminated at a Puffin crossing shall be as follows—
> (a) the red pedestrian light signal and the red figure shall both convey to a pedestrian the warning that, in the interests of safety, he should not cross the carriageway; and
> (b) the steady green pedestrian light signal and the steady green figure shall both indicate to a pedestrian that he may cross the carriageway and that drivers may not cause vehicles to enter the crossing.

(2) The flashing green pedestrian light signal at a Pelican crossing shall convey—
 (a) to a pedestrian who is already on the crossing when the flashing green signal is first shown the information that he may continue to use the crossing and that, if he is on the carriageway or a central reservation within the limits of that crossing (but not if he is on a central reservation which lies between two crossings which form part of a system of staggered crossings) before any part of a vehicle has entered those limits, he has precedence over that vehicle within those limits; and
 (b) to a pedestrian who is not already on the crossing when the flashing green light is first shown the warning that he should not, in the interests of safety, start to cross the carriageway.

(3) Any audible signal emitted by any device for emitting audible signals provided in conjunction with the steady green pedestrian light signal or the green figure, and any tactile signal given by any device for making tactile signals similarly provided, shall convey to a pedestrian the same indication as the steady green pedestrian light signal or as the green figure as the case may be.

8.9.5 Movement of Traffic at Crossings

Section IV of the 1997 Regulations contains the relevant regulations which restrict the movement of vehicles at crossings. Regulations 18 to 26 state:

18. The driver of a vehicle shall not cause the vehicle or any part of it to stop within the limits of a crossing unless he is prevented from proceeding by circumstances beyond his control or it is necessary for him to stop to avoid injury or damage to persons or property.

19. No pedestrian shall remain on the carriageway within the limits of a crossing longer than is necessary for that pedestrian to pass over the crossing with reasonable despatch.

20.—(1) For the purposes of this regulation and regulations 21 and 22 the word 'vehicle' shall not include a pedal bicycle not having a sidecar attached to it, whether or not additional means of propulsion by mechanical power are attached to the bicycle.

(2) Except as provided in regulations 21 and 22 the driver of a vehicle shall not cause it or any part of it to stop in a controlled area.

21. Regulation 20 does not prohibit the driver of a vehicle from stopping it in a controlled area—
 (a) if the driver has stopped it for the purpose of complying with regulation 25 or 26;
 (b) if the driver is prevented from proceeding by circumstances beyond his control or it is necessary for him to stop to avoid injury or damage to persons or property; or
 (c) when the vehicle is being used for police, fire brigade or ambulance purposes.

22.—(1) Regulation 20 does not prohibit the driver of a vehicle from stopping it in a controlled area—
 (a) for so long as may be necessary to enable the vehicle to be used for the purposes of—
  (i) any building operation, demolition or excavation;
  (ii) the removal of any obstruction to traffic;
  (iii) the maintenance, improvement or reconstruction of a road; or
  (iv) the laying, erection, alteration, repair or cleaning in or near the crossing of any sewer or of any main, pipe or apparatus for the supply of gas, water or electricity, or of any telecommunications apparatus kept installed for the purposes of a telecommunications code system or of any other telecommunications apparatus lawfully kept installed in any position,
 but only if the vehicle cannot be used for one of those purposes without stopping in the controlled area; or
 (b) if the vehicle is a public service vehicle being used—
  (i) in the provision of a local service; or
  (ii) to carry passengers for hire or reward at separate fares,
  and the vehicle, having proceeded past the crossing to which the controlled area relates, is waiting in that area in order to take up or set down passengers; or
 (c) if he stops the vehicle for the purpose of making a left or right turn.

(2) In paragraph (1) 'local service' has the meaning given in section 2 of the Transport Act 1985 but does not include an excursion or tour as defined by section 137(1) of that Act.

23. When vehicular light signals at a Pelican or Puffin crossing are displaying the red light signal the driver of a vehicle shall not cause it to contravene the prohibition given by that signal by virtue of regulation 12 or 13.

24.—(1) Whilst any motor vehicle (in this regulation called 'the approaching vehicle') or any part of it is within the limits of a controlled area and is proceeding towards the crossing, the driver of the vehicle shall not cause it or any part of it—

(a) to pass ahead of the foremost part of any other motor vehicle proceeding in the same direction; or

(b) to pass ahead of the foremost part of a vehicle which is stationary for the purpose of complying with regulation 23, 25 or 26.

(2) In paragraph (1)—

(a) the reference to a motor vehicle in sub-paragraph (a) is, in a case where more than one motor vehicle is proceeding in the same direction as the approaching vehicle in a controlled area, a reference to the motor vehicle nearest to the crossing; and

(b) the reference to a stationary vehicle is, in a case where more than one vehicle is stationary in a controlled area for the purpose of complying with regulation 23, 25 or 26, a reference to the stationary vehicle nearest the crossing.

25.—(1) Every pedestrian, if he is on the carriageway within the limits of a Zebra crossing, which is not for the time being controlled by a constable in uniform or traffic warden, before any part of a vehicle has entered those limits, shall have precedence within those limits, over that vehicle and the driver of the vehicle shall accord such precedence to any such pedestrian.

(2) Where there is a refuge for pedestrians or central reservation on a Zebra crossing, the parts of the crossing situated on each side of the refuge for pedestrians or central reservation shall, for the purposes of this regulation, be treated as separate crossings.

26. When the vehicular light signals at a Pelican crossing are showing the flashing amber signal, every pedestrian, if he is on the carriageway or a central reservation within the limits of the crossing (but not if he is on a central reservation which forms part of a system of staggered crossings) before any part of a vehicle has entered those limits, shall have precedence within those limits over that vehicle and the driver of the vehicle shall accord such precedence to any such pedestrian.

KEYNOTE

The expression 'pedestrian' replaced the former one of 'foot passengers' which had given rise to many debates about people pushing bicycles and skateboarders, rollerbladers, etc. It remains to be seen how the expression will be interpreted in relation to crossings.

Failing to observe the 1997 Regulations will amount to an offence under s. 25 of the 1984 Act above.

If a police officer, traffic warden or other authorised person is directing traffic at a crossing, it will for the time being cease to be 'uncontrolled' and their directions must be followed, even if they contravene the regulations. Failure to follow such directions may amount to an offence (**see chapter 10**).

Part I of sch. 1 to the 1997 Regulations requires Zebra crossings to be marked with yellow globes mounted at or near each end of the crossing. The globes must show a flashing or—where so authorised—a constant light (para. 1). If the globes have been disfigured, discoloured or have failed to light, the crossing will be still deemed to be properly marked for the purposes of enforcement (para. 4).

The globes may be fitted with boards to increase their visibility or prevent their light from reaching adjoining properties (para. 3).

The 'limits' of a Zebra crossing will be defined by the line of studs on each side of the black and white stripes. Anything or anyone who is outside those lines of studs is consequently outside the limits of the crossing (*Moulder* v *Neville* [1974] RTR 53).

Although the failure of a driver to stop at a crossing in contravention of the Regulations is *absolute* (that is, you need not show any particular state of mind, **see Crime, chapter 1**), if control of the vehicle is temporarily and unavoidably taken from the driver—say, by being shunted from behind by another vehicle—the driver will not commit an offence (*Burns* v *Bidder* [1966] 3 All ER 29).

Failure to observe crossing regulations is not of itself conclusive proof of driving without due care or consideration (*Gibbons* v *Kahl* [1956] 1 QB 59) (**see para. 2.4**).

8.9.6 School Crossings

The Road Traffic Regulation Act 1984 makes provision for the patrolling of school crossings (ss. 26–29). Section 28 states:

(1) When a vehicle is approaching a place in a road where a person is crossing or seeking to cross the road, a school crossing patrol wearing a uniform approved by the Secretary of State shall have power, by exhibiting a prescribed sign, to require the person driving or propelling the vehicle to stop it.

(2) When a person has been required under subsection (1) above to stop a vehicle—
- (a) he shall cause the vehicle to stop before reaching the place where a person is crossing or seeking to cross and so as not to stop or impede his crossing, and
- (b) the vehicle shall not be put in motion again so as to reach the place in question so long as the sign continues to be exhibited.

OFFENCE: **Failing to Comply—*Road Traffic Regulation Act 1984, s. 28(3)***

- Triable summarily • Fine • Discretionary disqualification

(No specific power of arrest)

The Road Traffic Regulation Act 1984, s. 28 states:

(3) A person who fails to comply with paragraph (a) of subsection (2) above, or who causes a vehicle to be put in motion in contravention of paragraph (b) of that subsection, shall be guilty of an offence.

KEYNOTE

This legislation has been amended by the Transport Act 2000 which removed the restriction on the times of day when a school crossing patrol could be operated. The former restrictions of 8.00 am to 5.30 pm have now been removed, enabling such crossings to be operated at any time. Furthermore, school crossing patrols are no longer restricted to stopping traffic when *children* are crossing and the newly-worded s. 28 enables patrols to stop traffic in order to allow anyone to cross safely.

This section creates two separate offences:

- s. 28(2)(a)—failing to stop in a way which does not impede the people crossing; and
- s. 28(2)(b)—*having stopped*, putting the vehicle in motion again while the sign continues to be exhibited.

Section 28 goes on to state:

(4) In this section—
- (a) 'prescribed sign' means a sign of a size, colour and type prescribed by regulations made by the Secretary of State or, if authorisation is given by the Secretary of State for the use of signs of a description not so prescribed, a sign of that description;
- (b) 'school crossing patrol' means a person authorised to patrol in accordance with arrangements under section 26 of this Act;

and regulations under paragraph (a) above may provide for the attachment of reflectors to signs or for the illumination of signs.

(5) For the purposes of this section—
- (a) where it is proved that a sign was exhibited by a school crossing patrol, it shall be presumed, unless the contrary is proved, to be of a size, colour and type prescribed, or of a description authorised, under subsection (4)(b) above, and, if it was exhibited in circumstances in which it was required by the regulations to be illuminated, to have been illuminated in the prescribed manner; and
- (b) where it is proved that a school crossing patrol was wearing a uniform, the uniform shall be presumed, unless the contrary is proved, to be a uniform approved by the Secretary of State.

KEYNOTE

The dimensions and details of the crossing sign exhibited by crossing patrol staff are set out in the School Crossing Patrol Sign (England and Wales) Regulations 2002 (SI 2002/3020). School crossings may also be marked with warning lights and these are provided for under the Traffic Signs Regulations and General Directions 2002 (SI 2002/3113).

Traffic wardens may act as school crossing patrols under the Functions of Traffic Wardens Order 1970 (SI 1970/1958) and when they do there is no need for them to wear the prescribed uniform (although they must display the approved sign).

8.10 Playgrounds

Section 29 of the Road Traffic Regulation Act 1984 makes provision for local traffic authorities to create orders prohibiting or restricting vehicular access to certain roads so that the roads may be used as 'street playgrounds'.

OFFENCE: **Contravention of Street Playground Order—*Road Traffic Regulation Act 1984, s. 29(3)***

- Triable summarily • Fine • Discretionary disqualification

(No specific power of arrest)

The Road Traffic Regulation Act 1984, s. 29 states:

(3) A person who uses a vehicle, or causes or permits a vehicle to be used, in contravention of an order in force under this section shall be guilty of an offence.

KEYNOTE

This offence does not apply to vehicles or drivers in the public service of the Crown.

For a discussion of the meaning of 'use, cause or permit', see chapter 1.

9 Construction and use

9.1 Introduction

The law governing the construction and use of road vehicles is mainly to be found—as the name suggests—in the Road Vehicles (Construction and Use) Regulations 1986 (SI 1986/1078). The Regulations deal with everything from motor cycle sidestands (reg. 38) to the placing of mascots (reg. 53). These Regulations are frequently amended and continue to be amended as the law develops. (For full details of the current Regulations, see *The Encyclopedia of Road Traffic Law & Practice* (Sweet & Maxwell).)

Although 'vehicles' for many purposes, pedal cycles have their own construction and use regulations (**see chapter 15**).

Lights on vehicles are dealt with at **para. 9.5** below.

Helpfully, the Road Vehicle (Construction and Use) Regulations 1986 have been drafted in two key parts. Part I deals with the construction, weight and equipment of vehicles (including trailers), while the second concentrates on the way in which those vehicles are used on roads. Invariably, all the features with which a vehicle must be fitted (brakes, wipers, warning instruments, etc.) can be found under part I of the Regulations, while the need to keep those features in good working order, other issues of vehicle maintenance and use on roads are addressed in part II.

So far as the *use* of road vehicles is concerned, in addition to the offences and safety measures outlined elsewhere in this work, the Road Traffic Reduction (National Targets) Act 1998 makes provision for the Secretary of State to set reduction targets for road traffic. Section 2 of the Act imposes a duty on the Secretary of State to set such targets in order to reduce the adverse effects of road traffic and to publish those targets. In discharging those duties the Secretary of State must have regard to a number of specified issues including:

- congestion
- danger to other road users
- the social impact of traffic levels
- the needs of people with disabilities, and
- the needs of rural communities.

For the purposes of the 1998 Act, road traffic generally means mechanically propelled vehicles using roads, with the exception of vehicles constructed or adapted to carry more than eight passengers in addition to the driver. Although proposals to introduce significant legislative changes in the regulation of road traffic were initially put on hold, the above Act came into force in July 1998 and there is no doubt as to the Government's intention to introduce further measures to regulate road usage in the UK.

For the specific duties of the traffic authority in relation to London, see part IV of the Greater London Authority Act 1999.

9.2 Vehicle Defect Rectification Scheme

The Vehicle Defect Rectification Scheme (VDRS) was introduced to streamline the process for dealing with vehicles having minor defects and to ensure that those defects were in fact rectified rather than simply punished. The scheme is voluntary and a motorist does not have to participate. The VDRS, which represents the next non-prosecution stage after a verbal warning, involves the person responsible being issued with a form by a police officer setting out the relevant defect in his/her vehicle. Once the defect has been pointed out and the person has accepted the form, he/she must have the defect remedied and then submit the vehicle for examination at a Department of Transport approved testing station. The testing station will certify that the defect has been rectified and the form, so endorsed, must be returned to the central police ticket office by the responsible person within 14 days. Failure to return the form within that time will result in the matter being considered for prosecution in the normal manner. A copy of the form is returned to the originating officer after 21 days.

9.3 Type Approval

Part I of the Regulations, which govern the way in which vehicles are built and equipped, are gradually being replaced by the 'type approval' system.

The type approval system has come about as a result of the United Kingdom's membership of the European Union whose common transport policy has had a significant impact on this area of law (see the Motor Vehicles (EC Type Approval) Regulations 1992 (SI 1992/3107), as amended).

The eventual aim of the policy is standardisation of the construction of all vehicles and their component parts.

When dealing with any potential breach of the legislation in this area, it is important to make sure whether a type approval system applies. In doing so you should refer to reg. 6 which contains guidance on the compliance with Community Directives and EC Regulations.

Whether or not a particular type approval scheme applies to a vehicle or a part used in one will largely be determined by the date of its first use or of its manufacture. It is important to distinguish which of these applies in any particular case.

The law which currently regulates this area is to be found in ss. 54–65 of the Road Traffic Act 1988 which makes it an offence to use, cause or permit to be used, a vehicle subject to the type approval requirements on a road without a certificate of conformity (s. 63(1)).

Additionally, under s. 80 of the 1988 Act the Secretary of State may designate the marking of motor vehicle parts to indicate conformity with a type approved by any country. Examples of such 'type approval marks' can be found under the Motor Vehicles (Designation of Approval Marks) Regulations 1979 (SI 1979/1088), as amended, which relate to braking systems and seat belt anchorage systems. Applying a 'mark' in a way that is likely to deceive is an offence under the Trade Descriptions Act 1968, ss. 24 and 25.

9.4 The Road Vehicles (Construction and Use) Regulations 1986

Regulation 3 sets out the relevant definitions; where there is no specific definition, those used under the Road Traffic Act 1988 will usually apply (**see chapter 1**).

In bringing a prosecution for an offence under these Regulations it is important to establish whether the correct offence is 'using', 'causing' or 'permitting'. These expressions are explained in **chapter 1**.

It is also important to establish whether a particular regulation applies to vehicles *first used* on or after a certain date; *first registered* on or after a certain date; or *manufactured* on or after a certain date (see e.g. *Mackinnon* v *Peate* [1936] 2 All ER 240).

Certain vehicles are exempted from the Regulations (see reg. 4) and again it is important to check that a vehicle in a particular case does fall within those categories.

9.4.1 Brakes

Regulations 15–18 set out the requirements as to braking systems on vehicles, together with those for their maintenance. It is not absolutely *necessary* for the person testing the braking system of a vehicle to be a 'qualified examiner' (see *Stoneley* v *Richardson* [1973] RTR 229 where a constable testified to being able to push the defendant's car along with the hand-brake applied).

The offence for breaching the requirement in relation to brakes is prosecuted under s. 41A of the Road Traffic Act 1988 (**see para. 9.4.10**).

See also s. 75 of the Road Traffic Act 1988 for the offence of selling unroadworthy vehicles.

9.4.2 Tyres

Section 41A of the 1988 Act also applies where the requirements in relation to tyres have been breached.

Regulation 24 sets out the requirements as to what tyres must be fitted to which vehicles, while reg. 25 restricts the speed limits and loads for such vehicles. Regulation 27 sets out a number of specific defects that will make tyres unlawful. It also contains exemptions for certain vehicles. The defects, in reg. 27, are where:

(1) ...

(a) the tyre is unsuitable having regard to the use to which the motor vehicle or trailer is being put or to the types of tyres fitted to its other wheels;

(b) the tyre is not so inflated as to make it fit for the use to which the motor vehicle or trailer is being put;

(c) the tyre has a cut in excess of 25 mm or 10% of the section width of the tyre, whichever is the greater, measured in any direction on the outside of the tyre and deep enough to reach the ply or cord;

(d) the tyre has any lump, bulge or tear caused by separation or partial failure of its structure;

(e) the tyre has any of the ply or cord exposed;

(f) the base of any groove which showed in the original tread pattern of the tyre is not clearly visible;

(g) either—

(i) the grooves of the tread pattern of the tyre do not have a depth of at least 1 mm throughout a continuous band measuring at least three-quarters of the breadth of the tread and round the entire outer circumference of the tyre; or

(ii) if the grooves of the original tread pattern of the tyre did not extend beyond three-quarters of the breadth of the tread, any groove which showed in the original tread pattern does not have a depth of at least 1 mm; or

(h) the tyre is not maintained in such condition as to be fit for the use to which the vehicle or trailer is being put or has a defect which might in any way cause damage to the surface of the road or damage to persons on or in the vehicle or to other persons using the road.

Regulation 27 also places restrictions on 're-cut' tyres (reg. 27(5)) and provides for many exemptions from the above conditions.

KEYNOTE

The entire outer circumference of a tyre does not usually include the outer walls or shoulder as they are not in contact with road (see *Coote* v *Parkin* [1977] RTR 61).
In the case of:

- passenger motor cars other than cars constructed or adapted to carry more than eight passengers (in addition to the driver)
- goods vehicles with a maximum gross weight not exceeding 3,500 kg
- light trailers

(first used on or after 3 January 1933 in each case)

the depth of tread requirement is increased to 1.6 mm throughout a continuous band across the central 3/4 section of the tyre and around the entire circumference (see the Road Vehicles (Construction and Use) (Amendment) (No. 4) Regulations 1990 (SI 1990/1981)).

To avoid the defect at reg. 27(1)(b) the tyre must be inflated so as to make it fit for the use to which the vehicle is being put at the material time; it does not have to be so inflated as to make it fit for some *future* use, however probable that use might be (*Connor* v *Graham* [1981] RTR 291).

Evidence as to the defective condition of a tyre may be given by anyone who saw it and it is no defence to argue that the tyre was not examined by an authorised vehicle examiner (*Phillips* v *Thomas* [1974] RTR 28).

Regulation 26 generally prohibits the mixing of different types of tyre (e.g. diagonal ply, bias-belted or radial ply) on both the same axle or different axles (although some combination is permissible). The use on roads of some 'knobbly' tyres or others which are designed for off-road use (on vehicles such as quad-bikes) may be prosecuted under reg. 27(1)(a) if it can be shown that the tyres were 'unsuitable having regard to the use to which the vehicle was put'.

'Tread pattern' includes plain surfaces as well as cut grooves but it does *not* cover tie-bars and tread wear indicators as used on many goods and heavy vehicles.

In relation to the selling of unsafe tyres, see the Motor Vehicles Tyres (Safety) Regulations 1994 (SI 1994/3117) and the Consumer Protection Act 1987, s. 12(5).

Exemptions

Regulation 27(4) sets out a number of exemptions to the provisions at (a)–(g) above, the main ones being:

- agricultural motor vehicles driven at not more than 20 mph
- agricultural trailers
- broken down vehicles or vehicles proceeding to a place to be broken up, being drawn in either case, by a motor vehicle at not more than 20 mph.

The onus appears to be on the defendant to show that his/her vehicle falls into an exempted category and that burden is not discharged simply by showing that the vehicle's excise licence describes it as being in one such category (*Wakeman* v *Catlow* [1977] RTR 174).

9.4.3 Mirrors

Regulation 33 sets out the requirements for mirrors on vehicles, including exemptions for certain types of vehicle. At reg. 33(1) there is a complex table showing the various types of mirror(s) required by different vehicles. Regulation 33(4) sets out the requirement for the stability and visibility of a vehicle's mirrors, together with the ability of the driver to adjust them.

9.4.4 Noise

Given the Government's commitment to exerting greater control over the effects that road traffic has on the community, these regulations in particular may become more prominent.

Regulations 54 and 57 regulate the fitting and use of silencers (including the rather sad attempts of some to increase the sound of their mopeds), while regs 37 and 99 apply to the types of audible warning instruments which are allowed on certain vehicles.

Regulation 97 provides that no motor vehicle shall be used on a road in such a manner as to cause any excessive noise which could have been avoided by the exercise of reasonable care by the driver.

A general prohibition on the use of an audible warning instrument while the vehicle is stationary on a road (except to warn of danger) is made by reg. 99, which also prohibits the use of such instruments between 11.30 pm and 7 am on a restricted road.

Following the introduction of 'boarding aid alarms' on certain public service vehicles (as to which, **see chapter 13**), these warning instruments have been added to regs 37 and 99. Even though they amount to audible warning instruments, boarding aid alarms may generally be sounded while the relevant vehicle is stationary (in fact, that's the whole point of them).

9.4.5 Quitting

Regulation 107 prohibits the leaving of a motor vehicle unattended on a road unless the engine has been stopped *and* the brake set; both must be done (*Butterworth* v *Shorthose* [1956] Crim LR 341). Any person left 'attending' the vehicle must be someone who is licensed to drive it and in a position to intervene otherwise reg. 107 is breached. This sort of behaviour is very common (particularly outside cashpoints or video rental outlets) and, owing to the danger it presents, can often be dealt with under s. 22 of the Road Traffic Act 1988 (**see chapter 8**).

For the duties of a driver to report an accident caused while a vehicle has been left in this way, **see chapter 4**.

There are some exceptions to this prohibition when the vehicle is being used for certain purposes (e.g. for police, ambulance or fire services).

9.4.6 Stopping of Engine

Regulation 98 provides that the driver of a vehicle when it is stationary shall stop the action of any machinery attached to or forming part of the vehicle so far as may be necessary for the prevention of noise *or exhaust emissions*. The italicised addition was inserted by the Road Vehicles (Construction and Use) (Amendment) Regulations 1998 (SI 1998 No. 1) and has significant implications for vehicles such as taxis waiting at ranks. The exceptions to reg. 98 are:

- when the vehicle is stationary owing to the necessities of traffic;
- where it is necessary to examine machinery following its failure or derangement, or where it is required to be worked for a purpose other than driving the vehicle; or
- where a vehicle is propelled by gas produced in plant carried on the vehicle, the exception applies to that plant itself.

9.4.7 Dangerous Use or Condition

Regulation 100 contains a catch-all provision to prevent the use of vehicles in a way which presents a danger to others. There are many ways in which that danger can be brought about (e.g. insecure loading, using the vehicle in a way which causes a nuisance, having the vehicle in a poor general condition). This regulation, which would apply to vehicles carrying too many people or vehicles being used for an unsuitable purpose, overlaps to a large extent with the offences under ss. 22A (**see chapter 8**) and 40A (**see para. 9.4.10**) of the Road Traffic Act 1988.

Regulation 104 makes provision to ensure that drivers maintain proper control of their vehicles and maintain a full view of the traffic ahead together with restrictions on the distances which drivers can reverse.

9.4.8 Reversing

Regulation 106 prohibits the reversing of a motor vehicle on a road further than may be requisite for the safety or reasonable convenience of the occupants or of other traffic (unless the vehicle is a road roller engaged in the construction, maintenance or repair of the road).

9.4.9 Special Types of Vehicle

The much amended Motor Vehicles (Authorisation of Special Types) General Order 1979 (SI 1979/1198) exempts a diverse collection of vehicles from the provisions of the 1986 Regulations.

Reading like an old Pepsi advert, the list of exemptions includes some track-laying, hedge-trimming, grass-cutting, life-saving, rotavating and excavating vehicles, together with some military and pedestrian-controlled vehicles. In the case of any vehicle which falls outside the conventional classes of cars, motor cycle, lorries and buses in terms of its size, construction or use, it is worth checking these exemptions.

The 1979 General Order also defines an *abnormal indivisible load.*

9.4.10 Offences

OFFENCE: **Using a Vehicle in Dangerous Condition etc.—*Road Traffic Act 1988, s. 40A***

- Triable summarily • Fine • Discretionary disqualification

(No specific power of arrest)

The Road Traffic Act 1988, s. 40A states:

> A person is guilty of an offence if he uses, or causes or permits another to use, a motor vehicle or trailer on a road when—
>
> (a) the condition of the motor vehicle or trailer, or of its accessories or equipment, or
> (b) the purpose for which it is used, or
> (c) the number of passengers carried by it, or the manner in which they are carried, or
> (d) the weight, position or distribution of its load, or the manner in which it is secured,
>
> is such that the use of the motor vehicle or trailer involves a danger of injury to any person.

KEYNOTE

Section 40A creates an offence of:

- using, causing or permitting another to use
- on a road
- a motor vehicle or trailer
- which, for whatever reason, involves a danger of injury
- to any person.

The existence of any danger is a question of fact, although s. 40A has to be read in conjunction with reg. 100(1) of the Road Vehicles (Construction and Use) Regulations 1986 (*DPP* v *Potts* [2000] RTR 1).

Given the broad wording of the offence, it would seem to invite an objective test of the use of the motor vehicle or trailer and whether or not that did in fact create a potential danger to anyone including the driver. However, the offence applies to *motor* vehicles and trailers only and not to any other type of vehicle (for the respective definitions of these terms, see chapter 1).

The test in relation to the potential for injury is an objective one and will consider the anticipated eventualities of the ordinary course of driving including the need for sudden braking, swerving, etc. (*Akelis* v *Normand* 1997 SLT 136). In a recent case, a seven year old boy was seen to be travelling in the open back of an uncovered jeep without any fitted restraints. The boy was steadying himself by holding on to the vehicle's roll-bars. The court held that, even though he had travelled in that way without incident many times in the past and that his father, the driver, was generally a responsible parent, the objective test as to the potential for injury meant that the offence (under s. 40A(c) above) had been committed (*Gray* v *DPP* [1999] RTR 339).

OFFENCE: **Breach of Requirement: Brakes, Steering Gear or Tyres—*Road Traffic Act 1988, s. 41A***

- Triable summarily • Fine • Discretionary disqualification under specified conditions

(No specific power of arrest)

The Road Traffic Act 1988, s. 41A states:

A person who—
(a) contravenes or fails to comply with a construction and use requirement as to brakes, steering-gear or tyres, or
(b) uses on a road a motor vehicle or trailer which does not comply with such a requirement, or causes or permits a motor vehicle or trailer to be so used,

is guilty of an offence.

OFFENCE: **Breach of Weight Requirements—*Goods and Passenger Vehicles—Road Traffic Act 1988, s. 41B***

- Triable summarily • Fine

(No specific power of arrest)

The Road Traffic Act 1988, s. 41B states:

(1) A person who—
    (a) contravenes or fails to comply with a construction and use requirement as to any description of weight applicable to—
        (i) a goods vehicle, or
        (ii) a motor vehicle or trailer adapted to carry more than eight passengers, or
    (b) uses on a road a vehicle which does not comply with such a requirement, or causes or permits a vehicle to be so used,
    is guilty of an offence.

Defence

The Road Traffic Act 1988, s. 41B states:

(2) In any proceedings for an offence under this section in which there is alleged a contravention of or failure to comply with a construction and use requirement as to any description of weight applicable to a goods vehicle, it shall be a defence to prove either—
    (a) that at the time when the vehicle was being used on the road—
        (i) it was proceeding to a weighbridge which was the nearest available one to the place where the loading of the vehicle was completed for the purpose of being weighed, or
        (ii) it was proceeding from a weighbridge after being weighed to the nearest point at which it was reasonably practicable to reduce the weight to the relevant limit, without causing an obstruction on any road, or
    (b) in a case where the limit of that weight was not exceeded by more than 5 per cent—
        (i) that that limit was not exceeded at the time when the loading of the vehicle was originally completed, and
        (ii) that since that time no person has made any addition to the load.

OFFENCE: **Breach of Other Construction and Use Requirements—*Road Traffic Act 1988, s. 42***

• Triable summarily. • Fine.

(No specific power of arrest)

The Road Traffic Act 1988, s. 42 states:

A person who—

(a) contravenes or fails to comply with any construction or use requirement other than one within section 41A(a) or 41B(1)(a) of this Act, or

(b) uses on a road a motor vehicle or trailer which does not comply with such a requirement, or causes or permits a motor vehicle or trailer to be so used,

is guilty of an offence.

9.5 Lights

The law which governs the fitting and use of lights on vehicles is to be found in the Road Vehicles Lighting Regulations 1989 (SI 1989/1796). They are made under s. 41 of the Road Traffic Act 1988, for all vehicles except cycles (which are made under s. 81).

Parts of the 1989 Regulations are extremely detailed and reference should be made to the specific wording or diagram in each case. This chapter addresses only the broader aspects of the lighting regulations.

As with the Road Vehicles (Construction and Use) Regulations 1986 above, the lighting regulations can be divided into regulations which address fitting, maintenance and use.

The main lights are themselves divided into:

- headlamps (main beam and dipped)
- front and rear position lamps (side lights)
- front and rear fog lamps
- stop lamps
- reversing lamps
- optional lamps
- rear registration plate lamps.

Also included in the 1989 Regulations are reflectors and markers.

9.5.1 Offences

Offences contravening the lighting regulations are charged under s. 41 of the Road Traffic Act 1988 in the same ways as the other construction and use offences above.

Note the wording of a prohibition under the relevant regulation may be aimed at the fitting of a light and not necessarily its use.

9.5.2 Definitions

Regulation 3 sets out the definitions for the purposes of the 1989 Regulations. Again, where none is specified, the corresponding definition under the Road Traffic Act 1988 will usually apply.

The Regulations separate a 24-hour period into:

- the period between sunset and sunrise
- the period between sunrise and sunset.

Where the old distinction of 'daytime hours' and 'hours of darkness' apply, reg. 3 defines them as 'the time between half an hour before sunrise and half an hour after sunset' and 'the time between half an hour after sunset and half an hour before sunrise' respectively.

9.5.3 Exemptions

Regulations 4–9A list the exemptions which include:

4. ...

(3) Nothing in these Regulations shall require any lamp or reflector to be fitted between sunrise and sunset to—
- (a) a vehicle not fitted with any front or rear position lamp,
- (b) an incomplete vehicle proceeding to a works for completion,
- (c) a pedal cycle,
- (d) a pedestrian-controlled vehicle,
- (e) a horse-drawn vehicle,
- (f) a vehicle drawn or propelled by hand, or
- (g) a combat vehicle.

5. [Temporarily imported vehicles and vehicles proceeding to a port for export]

6. [Vehicles towing or being towed]

7. [Military vehicles]

8. [Invalid carriages]

9. [Vehicles drawn or propelled by hand]

9A. [Tram cars]

Other exemptions will apply to some parked vehicles (see e.g. reg. 24).

9.5.4 Fitting of Lights

Regulations 11–22 cover the fitting of lights, reflectors and rear markings. The vehicles shown in sch. 1 must be fitted with the corresponding lights etc. and it is an offence to use, or to cause or permit to be used (**see chapter 1**) such a vehicle which does not comply with those regulations.

Regulation 11 deals with the colours of lights and reflectors which must or must not be fitted, while regs 13 and 16 address the fitting of flashing lights and warning beacons.

Regulations 21 and 22 deal with the fitting of additional lights to long vehicles and trailers.

9.5.5 Rear Registration Lamps

Although rear registration lamps are addressed under the 1989 Regulations, there is also an offence under the Vehicle Excise and Registration Act 1994 (**see chapter 12**).

9.5.6 Headlamps

Schedules 4 and 5 contain the specifications for headlamps.

Regulation 25 makes it an offence to use or cause or permit to be used, a vehicle with obligatory headlamps on a road unless they are kept lit during the hours of darkness and in seriously reduced visibility. (There is an exception to this requirement which applies to vehicles on 'restricted' roads per s. 81 of the Road Traffic Regulation Act 1984 where the street lights are on.)

9.5.7 Position Lamps

Schedules 2 and 10 set out the specification for front and rear position lamps respectively. Regulation 24 makes it an offence to use, or to cause or permit to be used, a vehicle on a road between sunset and sunrise or in motion in seriously reduced visibility, unless every position lamp, rear registration lamp and side-marker lamp required by it are lit.

9.5.8 Maintenance and Use

Regulations 23–27 govern the maintenance of lights and reflectors, together with their use. Regulation 23 makes it an offence to use, or to cause or permit to be used, a vehicle on a road unless the relevant lamps and reflectors (including some optional ones) are clean and in good working order.

The general exception to reg. 23 is if the vehicle is being used during daytime hours and the defect only happened during the journey or if arrangements have been made to rectify it with all reasonable expedition (reg. 23(3)).

Regulation 27 sets out a table showing the ways in which certain lights may/may not be used (**see appendix 5**). Regulation 27 states that no person shall use (or cause or permit to be used) on a road, any vehicle on which any lamp, hazard warning signal/device or beacon of a type specified in column 2 of the table in a manner specified in column 3.

9.5.9 Motor Cycles

A number of the construction and use regulations relate specifically to motor cycles.

Regulations 57 and 57A set out the construction requirements in relation to noise limits and exhaust systems on motor cycles.

Regulation 84 imposes restrictions on the drawing of trailers by motor cycles on a road, while regs 92 and 93 make provision for the attachment and use of side cars.

9.6 Testing

Section 47 of the Road Traffic Act 1988 requires all motor vehicles which were first registered more than three years before the time when they are being used on a road to pass a test. That requirement includes vehicles manufactured abroad (s. 47(2)(b)).

Some vehicles need to be tested after one year, notably:

- motor vehicles having more than eight seats (excluding the driver's seat) which are used to carry passengers
- taxis
- ambulances
- larger goods vehicles.

The procedure for testing of vehicles is set out in the Motor Vehicles (Tests) Regulations 1981 (SI 1981 No. 1694), as amended. These and their many amending Regulations lay down both the instructions for carrying out a test and also specify those items which will be tested.

Tests can only be carried out by 'authorised examiners' and others listed in s. 45(3) and vehicle examiners appointed under s. 66A (as amended by the Road Traffic (Vehicle Testing) Act 1999). Many garages are so authorised. Roadside tests may also be carried out in some circumstances, **see para. 9.6.2**.

Where a garage returned a car to its owner on the understanding that it had been repaired and had passed its MOT when in fact it was unroadworthy, the garage was held to have committed the offence of 'supplying' an unroadworthy vehicle contrary to s. 75 of the Road Traffic Act 1988 (*Devon County Council* v *DB Cars* [2002] Crim LR 71). In that case the High Court held that the word 'supply' involved no more than a transfer of physical control of an item from one person to another in order to provide the other person with something that he/she wanted. As the garage owner transferred physical control of the car to the owner he had, on that definition, 'supplied' it. (See also *Formula One Autocentres Ltd* v *Birmingham City Council* [1999] RTR 195.) This odd decision is not followed in other offences involving 'supply' such as those under the Misuse of Drugs Act 1971 (**see Crime, chapter 6**).

For the scope of Regulations that may be made in relation to the testing of vehicles, see the Road Traffic Act 1988, s. 46. The Road Traffic (Vehicle Testing) Act 1999 makes a number of significant changes to the statutory framework for the testing of vehicles. It will introduce a system to operate a national computerised database of vehicles and their respective issue/need for MOTs. The 1999 Act will allow for changes to be made in respect of a number of aspects including:

- the supervision of MOT testing centres
- the training of staff
- access to information held on the database
- the admissibility/use of records as an alternative to production of the test certificate itself.

The 1999 Act also amends the summary offence of impersonating an authorised examiner with intent to deceive under s. 177 of the Road Traffic Act 1988.

The items tested on a vehicle will include:

- brakes
- tyres
- seat belts
- steering
- certain lights/reflectors
- stop lights
- exhausts
- wipers and washers
- direction indicators
- suspension
- bodywork
- horn.

Satisfactory completion of the test will result in the issuing of a certificate under s. 45(2)(b).

The Motor Vehicles (Tests) (Amendment) Regulations 1998 (SI 1998/1672) introduced new transitory classes of vehicle to allow for extra installation checks to be made to minibuses, buses and coaches in respect of their seat belts.

OFFENCE: **Using, Causing or Permitting Use of Vehicle without Test Certificate—*Road Traffic Act 1988, s. 47(1)***

- Triable summarily • Fine

(No specific power of arrest)

The Road Traffic Act 1988, s. 47 states:

(1) A person who uses on a road at any time, or causes or permits to be so used, a motor vehicle to which this section applies, and as respects which no test certificate has been issued within the appropriate period before that time, is guilty of an offence.
In this section and section 48 of this Act, the 'appropriate period' means a period of twelve months or such shorter period as may be prescribed.

KEYNOTE

As ever, there are numerous exemptions to the requirements of s. 47, most of which relate to larger vehicles, track-laying vehicles and some pedestrian controlled vehicles (see reg. 6 of the 1981 Regulations).

Some military vehicles, some electrically-powered goods vehicles and vehicles temporarily in Great Britain are also exempt.

Vehicles *provided* (as opposed to 'used') for police purposes are exempt if they are maintained in an approved police workshop (reg. 6(xiv)). Other exemptions exist in relation to vehicles seized or detained by the police or customs and excise officers.

Special provision is made (under s. 48) for the issue of temporary exemption certificates in the case of certain public service vehicles.

An exemption exists where the person using the vehicle is taking it to or from a testing centre. This exemption is only applicable where the test has been previously arranged with the garage. A further exemption exists where a test certificate has been refused and the vehicle is:

- being delivered by prior arrangement, or brought from the place where the relevant work is to be/has been carried out; or
- being towed to a place where it is to be broken up.

Provision is made in reg. 6 for the use of a vehicle by an authorised examiner or inspector during the test.

The provisions of s. 165 of the Road Traffic Act 1988 apply to test certificates (see para. 6.2.6).

This means that an officer may issue a form HO/RT 1 in respect of such a certificate.

Note that the offences under s. 173 (forging documents) and s. 175 (issuing false documents) of the Road Traffic Act 1988 apply to test certificates (see chapter 16).

9.6.1 Goods Vehicles

The testing of goods vehicles is governed by ss. 49–53 of the Road Traffic Act 1988 and the Goods Vehicles (Plating and Testing) Regulations 1988 (SI 1988/1478), as amended.

They are tested and 'plated' at government testing stations under the ambit of the Secretary of State and these tests must be carried out annually.

Schedule 2 to the Regulations lists the many vehicles which are exempt, a list which includes snow ploughs, hearses, fire fighting vehicles and road rollers.

OFFENCE: **Using, Causing or Permitting Use of Goods Vehicle without Test Certificate—*Road Traffic Act 1988, s. 53(2)***

- Triable summarily • Fine

(No specific power of arrest)

The Road Traffic Act 1988, s. 53 states:

(2) If any person at any time on or after the relevant date—

(a) uses on a road a goods vehicle of a class required by regulations under section 49 of this Act to have been submitted for a goods vehicle test, or

(b) causes or permits to be used on a road a goods vehicle of such a class,

and at that time there is no goods vehicle test certificate in force for the vehicle, he is guilty of an offence.

In this subsection 'relevant date', in relation to any goods vehicle, means the date by which it is required by the regulations to be submitted for its first goods vehicle test.

Section 53(1) creates a similar offence in relation to Plating Certificates.

9.6.2 Roadside Tests

Section 67 of the Road Traffic Act 1988 allows 'authorised examiners' to carry out roadside tests on motor vehicles, in relation to the brakes, steering, tyres, lights and noise and fume emission. Authorised examiners include police constables so authorised by their chief officer of police. Other examiners may be appointed (e.g. by the Secretary of State or a police authority) but they must produce their authority to act as such if required to do so (s. 67(5)). Obstructing such an examiner is a summary offence under s. 67(9) (for the offence of obstructing a police officer generally, **see Crime, chapter 8**).

Section 67(6) allows for drivers to ask for the examination to be deferred (in accordance with the time limits set out at sch. 2 to the 1988 Act). However, s. 67 goes on to state:

(7) Where it appears to a constable that, by reason of an accident having occurred owing to the presence of the vehicle on a road, it is requisite that a test should be carried out forthwith, he may require it to be so carried out and, if he is not to carry it out himself, may require that the vehicle shall not be taken away until the test has been carried out.

(8) Where in the opinion of a constable the vehicle is apparently so defective that it ought not to be allowed to proceed without a test being carried out, he may require the test to be carried out forthwith.

(9) If a person obstructs an authorised examiner acting under this section, or fails to comply with a requirement of this section or Schedule 2 to this Act, he is guilty of an offence.

(10) In this section and in Schedule 2 to this Act—
(a) 'test' includes 'inspect' or 'inspection', as the case may require, and
(b) references to a vehicle include references to a trailer drawn by it.

KEYNOTE

For the 'power' to stop vehicles for these purposes, **see chapter 10**.

Section 68 provides a power for vehicle examiners to inspect goods vehicles, PSVs and some larger passenger carrying vehicles. It also provides a power for a police officer in uniform to direct such a vehicle to a suitable place of inspection when found on a road, provided that that place of inspection is not more than five miles from the place where the requirement is made. Refusal or neglect to comply is a summary offence. The power is only available in relation to vehicles that are 'stationary' on a road. Presumably if the vehicle is moving, the officer need only stop it (using the general power under s. 163) before directing it to be driven to a place of inspection.

9.6.3 Testing and Inspection

Regulation 74 of the Road Vehicles (Construction and Use) Regulations 1986 provides a power to test and inspect the:

- brakes
- silencers
- steering gear
- tyres

of any vehicle on any premises where that vehicle is located.

The power applies to police officers in uniform and other authorised vehicle examiners (see reg. 74(a)–(f)). Regulation 74 provides no power of entry and stipulates that the person empowered shall produce his/her authorisation if required to do so. It also provides that no such test or inspection shall be carried out unless:

- the owner of the vehicle consents;
- notice has been given to that owner (either personally or left at his/her address not less than 48 hours before the time of the proposed test/inspection, or sent to him/her by recorded delivery at least 72 hours before the proposed test/inspection); or
- the test or inspection is made within 48 hours of a reportable accident in which the vehicle was involved.

10 Traffic signs and directions

10.1 Introduction

A 'traffic sign' is defined under the Road Traffic Regulation Act 1984, s. 64 which states:

> In this Act 'traffic sign' means any object or device (whether fixed or portable) for conveying, to traffic on roads or any specified class of traffic, warnings, information, requirements, restrictions or prohibitions of any description—
> (a) specified by regulations made by the Ministers acting jointly; or
> (b) authorised by the Secretary of State,
> and any line or mark on a road for so conveying such warnings, information, requirements, restrictions or prohibitions.

Such a traffic sign will include temporary signs and road markings (e.g. double white lines).

The restrictions on prosecution imposed by s. 1 of the Road Traffic Offenders Act 1988 (notices of intended prosecution) should be borne in mind when considering offences in this chapter.

Note that, since the Road Traffic Act 1991, increasing reliance has been placed on the use of automatic devices for detecting traffic light offences and that the procedure of 'conditional offers' of fixed penalties is available for such offences (**see chapter 14**).

10.2 Authorised Traffic Signs

As a starting point, the specific dimensions, colours and other requirements for any traffic signs—including lights and other signals—will generally be found in the Traffic Signs Regulations and General Directions 2002 (SI 2002 No. 3113). This statutory instrument is in two parts, each of which can be cited by its own relevant title (e.g. the Traffic Signs Regulations or the Traffic Signs General Directions). Part I (the Regulations) and its many schedules deal with the prescribed detail and layout of signs, along with their placement. Part I (section 5) also deals with lights and warning signals, including portable signals. The technical specifications for pedestrian crossing facilities provided in conjunction with traffic light signals are provided for in reg. 47, while combined pedestrian and cyclist crossings (Toucan crossings) come under reg. 49. Part II (the General Directions) is created under s. 65 of the Road Traffic Regulation Act 1984 (**see para. 10.2.1**) and deals with the placing of signs and road markings. The whole statutory instrument makes provision for the Secretary of State and the National Assembly for Wales to authorise other traffic signs. Part I also makes it clear that this legislation does not authorise the placing of signs indicating a temporary obstruction (**see para. 10.2.3**). While there are some transitional arrangements in place and some signs prescribed by earlier regulations are adopted by the statutory instrument, the Traffic Signs Regulations and General Directions 2002 will cover most occasions involving the dimensions and positioning of traffic signs and signals.

If a traffic sign is to have any legal effect you must show that it was placed or displayed by someone who was authorised to do so under the circumstances.

Traffic signs, whether temporary or permanent, may not be placed on or near a road unless the person placing them is specifically authorised to do so (Road Traffic Regulation Act 1984, s. 64(4)).

Among those authorised are:

- **the local traffic authority**—in respect of roads in their area;
- **a constable or other person authorised by the chief officer of police**—under certain circumstances;
- **any other person**—in respect of a temporary obstruction.

10.2.1 Traffic Authority

Section 65 of the Road Traffic Regulation Act 1984 empowers a traffic authority to cause or permit traffic signs to be placed on or near any road in their area. These signs must be placed in accordance with certain directions issued by the Secretary of State and/or Minister, most of which fall under the Traffic Signs Regulations and General Directions 2002.

Special provisions exist for the placing of traffic signs in London under chapter XIII of the Greater London Authority Act 1999. Part IV of the 1999 Act creates a body corporate known as 'Transport for London' and sets out its many functions as the capital's traffic authority, including taxis and the London Underground.

The civil liability of a traffic authority when placing, or deciding not to place, traffic signs under this section must be judged against the House of Lords' decision in *Stovin* v *Wise* (*Norfolk CC, third party*) [1996] RTR 354. There it was held that there are only limited grounds for apportioning civil liability to a traffic authority where it has taken a decision in relation to its statutory powers. In exercising its powers under s. 65, a traffic authority owes a duty of care to motorists when selecting a suitable position for a particular traffic sign (*Levine* v *Morris* [1970] 1 WLR 71).

10.2.2 Constable or Other Person Authorised by Chief Officer of Police

Sections 66 and 67 of the Road Traffic Regulation Act 1984 provide powers for police officers (and others acting under the instructions of a chief officer) to place traffic signs. Section 66 applies to the placing of signs and 'structures' in relation to local traffic regulations and includes:

- Races or speed trials authorised under s. 31(4) of the Road Traffic Act 1988.
- Measures taken to avoid obstruction on public occasions or near public buildings in London under s. 52 of the Metropolitan Police Act 1839.
- Similar measures authorised under other statutes (e.g. the Town Police Clauses Act 1847 or enactments relating to the City of London).

Section 67 of the 1984 Act states:

> (1) A constable, or a person acting under the instructions (whether general or specific) of the chief officer of police, may place on a road, or on any structure on a road, traffic signs (of any size, colour and type prescribed or authorised under section 64 of this Act) indicating prohibitions, restrictions or requirements relating to vehicular traffic, as may be necessary or expedient to prevent or mitigate congestion or obstruction of traffic, or danger to or from traffic, in consequence of extraordinary circumstances; and the power to place signs conferred by this subsection shall include power to maintain a sign for a period of seven days or less from the time when it was placed, but no longer.

KEYNOTE

The offence of failing to comply with traffic directions (s. 36 of the Road Traffic Act 1988 (**see para. 10.3**) applies to signs placed under the authority granted by s. 67(1) (s. 67(2)).

10.2.3 Temporary Obstructions

The placing of traffic signs to warn of temporary obstructions is governed by the Traffic Signs (Temporary Obstructions) Regulations 1997 (SI 1997 No. 3053).

The 1997 Regulations introduced a new sign, the road vehicle sign, which is intended to warn of a temporary obstruction caused by stationary vehicles. They also make provision for someone in charge of, or accompanying emergency or breakdown vehicles, to place 'Keep Right' signs under certain circumstances.

Authorisation to Place Temporary Signs

Regulation 15 states:

(1) Subject to paragraph (4) of this regulation, a person who is in charge of or accompanies an emergency or a breakdown vehicle which is temporarily obstructing a road is hereby authorised to place a keep right sign for the purpose of warning vehicular traffic of the obstruction created by the vehicle and to indicate the way past the vehicle.

(2) Subject to paragraph (4) of this regulation, any person not otherwise authorised to do so is hereby authorised to place a road vehicle sign on a vehicle or a flat traffic delineator, traffic cone, traffic pyramid, traffic triangle or warning lamp on any road for the purpose of warning traffic of a temporary obstruction in the road, other than one caused by the carrying out of works.

KEYNOTE

Regulation 16 requires that signs be placed in an upright position and in such a way as to guide traffic past the obstruction. It also requires that they should be removed as soon as the relevant obstruction has itself been removed.

'Emergency vehicles' here will generally include vehicles used for fire, ambulance and police purposes, rescue vehicles and vehicles used in connection with blood transfusions and tissue transplant. The full definition can be found in the Road Vehicles Lighting Regulations 1989 (SI 1989 No. 1796).

'A breakdown vehicle' here is a vehicle used to attend an accident or a breakdown or to draw a broken down vehicle.

Temporary Traffic Signs

Part II of the 1997 Regulations states:

4.—(1) In addition to the requirement conveyed by the sign shown in diagram 610 of the 1994 Regulations in accordance with those Regulations, a keep right sign shall convey to vehicular traffic a warning of a temporary obstruction.

(2) Section 36 of the Road Traffic Act 1988 shall apply to the keep right sign.

5. In addition to indicating the edge of a route for vehicular traffic through or past a temporary obstruction, in accordance with the 1994 Regulations, a traffic cone and a flat delineator shall each convey to such traffic on a road a warning of an obstruction in the road.

6. A road vehicle sign shall convey to vehicular traffic using a road a warning of a temporary obstruction in the road caused by a stationary vehicle.

7. A traffic pyramid, a traffic triangle and a warning lamp shall each convey to vehicular traffic using a road a warning of a temporary obstruction in the road, other than an obstruction caused by the carrying out of works.

Part III sets out the requirements for the dimensions of certain traffic signs, including:

- 'keep right' signs
- flat traffic delineators
- road vehicle signs
- traffic cones, pyramids and triangles
- warning lamps.

Diagrams showing the requirements for road vehicle signs, pyramids and triangles can be found in schs 1–3 of the 1997 Regulations.

10.3 Failing to Comply with Traffic Sign

Depending on the sign involved, failure to comply with a traffic sign can, as a general rule, be divided into three broad categories:

1. Failing to comply with statutory restrictions or requirements (that is, under public or local Acts as opposed to Orders or Regulations). These cases are punishable under s. 36 of the Road Traffic Act 1988.
2. Failing to comply with signs where s. 36 of the Road Traffic Act 1988 is specifically referred to as applying to them (e.g. those listed in reg. 10 of the Traffic Signs Regulations and General Directions 2002).
3. Failing to comply with other 'authorised signs'. These cases are punishable under s. 91 of the Road Traffic Offenders Act 1988.

There is an overlapping category, namely failing to stop or observe a direction given by an authorised person engaged in directing traffic. For a full discussion of these, **see para. 10.4**.

Those offences where s. 36 of the Road Traffic Act 1988 applies are subject to the Notice of Intended Prosecution procedure (**see chapter 3**).

The following signs are among those listed in reg. 10 of the Traffic Signs Regulations and General Directions 2002:

- 'STOP' signs
- 'GIVE WAY' signs at junctions and open level crossings
- 'KEEP LEFT/RIGHT' signs
- NO 'U' TURN signs
- 'NO ENTRY' signs
- 'STOP' board, operated manually on lengths of road where one-way working is necessary
- signs prohibiting vehicles of a certain height
- signs indicating bus or pedal cycle route
- red traffic light signals
- green traffic light filter signals
- double white line systems.

OFFENCE: **Failing to Comply with a Traffic Sign—*Road Traffic Act 1988, s. 36(1)***

- Triable summarily • Fine
- Discretionary disqualification under certain circumstances

(No specific power of arrest)

The Road Traffic Act 1988, s. 36 states:

(1) Where a traffic sign, being a sign—
 (a) of the prescribed size, colour and type, or
 (b) of another character authorised by the Secretary of State under the provisions in that behalf of the Road Traffic Regulation Act 1984,

has been lawfully placed on or near a road, a person driving or propelling a vehicle who fails to comply with the indication given by the sign is guilty of an offence.

KEYNOTE

This offence only applies to failure to comply with those signs listed at categories 1. and 2. above. Regulation 10(2) of the Traffic Signs Regulations and General Directions 2002 sets out those traffic signs which, if contravened, will allow discretionary disqualification by a court. The regulation contains the full details of the signs referred to in column 5 of the entry in sch. 2 to the Road Traffic Offenders Act 1988 (**see appendix 3**).

Section 36(3) states:

(3) For the purposes of this section a traffic sign placed on or near a road shall be deemed—
 (a) to be of the prescribed size, colour and type, or of another character authorised by the Secretary of State under the provisions in that behalf of the Road Traffic Regulation Act 1984, and
 (b) (subject to subsection (2)...) to have been lawfully so placed

unless the contrary is proved.

OFFENCE: **Failing to Comply with Other Traffic Sign—*Road Traffic Offenders Act 1988, s. 91***

- Triable summarily • Fine

(No specific power of arrest)

The Road Traffic Offenders Act 1988, s. 91 states:

If a person acts in contravention of or fails to comply with—
 (a) any regulations made by the Secretary of State under the Road Traffic Act 1988 other than regulations made under section 31, 45 or 132,
 (b) any regulations made by the Secretary of State under the Road Traffic Regulation Act 1984, other than regulations made under section 28, Schedule 4, Part III of Schedule 9 or Schedule 12,

and the contravention or failure to comply is not made an offence under any other provision of the Traffic Acts, he shall for each offence be liable...

KEYNOTE

If the sign which a driver contravenes is not covered by s. 36 of the Road Traffic Act 1988 (e.g. failing to conform to an 'amber' traffic light), the appropriate charge would generally be under this Act and section.

10.4 Powers to Stop and Direct Traffic

10.4.1 Power to Stop Vehicles

Most of the police powers to stop vehicles and to direct traffic are statutory. However, as part of their overall common law powers (and duties) to protect life and property, the police have additional powers in this regard. These can include, under some very unusual circumstances, the power to require a motorist to carry out what would otherwise be an unlawful manoeuvre such as reversing a short distance up a one-way street the wrong way (see *Johnson* v *Phillips* [1975] 3 All ER 682).

Section 163(1) and (2) of the Road Traffic Act 1988 states:

> (1) A person driving a mechanically propelled vehicle on a road must stop the vehicle on being required to do so by a constable in uniform.
>
> (2) A person riding a cycle on a road must stop the cycle on being required to do so by a constable in uniform.

KEYNOTE

The above provisions give uniformed police officers their general powers to stop mechanically propelled vehicles on roads. There are other specific powers to stop vehicles under certain circumstances (**see General Police Duties, chapter 2**). This general power is extended to designated community support officers under sch. 4 to the Police Reform Act 2002 within their relevant police area but *only* for the purposes of carrying out road checks (**see General Police Duties, chapter 2**).

There is also a strangely worded provision in s. 67(3) of the Road Traffic Act 1988 relating to roadside vehicle tests (as to which, **see chapter 9**). Section 67(3) says that 'a vehicle shall not be required to stop for a test except by a constable in uniform'. Although this is not specifically expressed as a power (but rather a restriction on the use of any other powers to stop vehicles other than the general police power under s. 163), this provision is extended to traffic wardens (see the Functions of Traffic Wardens (Amendment) Order 2002 (SI 2002 No. 2975)) and to all intents and purposes is treated as a power. It is also extended to designated community support officers and accredited employees under schs 4 and 5 to the Police Reform Act 2002 (as to which, **see General Police Duties, chapter 2**).

OFFENCE: **Failing to Stop when Required—*Road Traffic Act 1988, s. 163(3)***

- Triable summarily • Fine

(No specific power of arrest)

The Road Traffic Act 1988, s. 163 states:

> If a person fails to comply with this section he is guilty of an offence.

KEYNOTE

The exercise of powers to carry out road checks under Part I of the Police and Criminal Evidence Act 1984 is covered by s. 163, that is, if a driver fails to stop when required by a constable in uniform at such a road check, he/she will commit this offence (for a discussion of this and other powers relating to the stopping of motor vehicles, **see General Police Duties, chapter 2**).

The lawful exercise of police powers to stop people going about their business within the community can be a significant source of friction. The manner or frequency of the use of power such as the power to stop vehicles on a road might be perceived by the community as a source of oppression and discrimination, leading to a reduction in confidence in the police and the creation of an atmosphere of distrust. As with stop and search powers generally (**see General Police Duties, chapter 2**), this was highlighted in the Stephen Lawrence inquiry (Cm 4262–I), see para. 46.31 and recommendation 61.

'Stop' means bringing the vehicle to a halt and remaining at rest for long enough for the officer (or traffic warden; **see para. 10.4.2**) to exercise whatever additional powers may be appropriate (see *Lodwick* v *Sanders* [1985] 1 All ER 577).

Unlike the power under s. 35(1) below, this power is confined to 'mechanically propelled vehicles' (**see chapter 1**) or cycles.

The power of entry under s. 17 of the Police and Criminal Evidence Act 1984 applies to this offence (**see General Police Duties, chapter 2**).

In addition, a constable in uniform may arrest a person without warrant if he/she has reasonable cause to suspect that the person has committed the above offence after 1 October 2002 (Road Traffic Act 1988, s. 163(4)).

Use of this power by police officers to detect drink/driving offences was endorsed in *Chief Constable of Gwent* v *Dash* [1986] RTR 41.

The Terrorism Act 2000 also gives the police specific powers to stop *any* vehicles in relation to authorised stop and search operations under certain conditions (**see chapter 8** and **General Police Duties, chapter 2**).

OFFENCE: **Failing to Comply with Traffic Directions of Constable—*Road Traffic Act 1988, s. 35(1)***

- Triable summarily • Fine

(No specific power of arrest)

The Road Traffic Act 1988, s. 35 states:

(1) Where a constable is for the time being engaged in the regulation of traffic in a road, a person driving or propelling a vehicle who neglects or refuses—
(a) to stop the vehicle, or
(b) to make it proceed in, or keep to, a particular lane of traffic,
when directed to do so by the constable in the execution of his duty is guilty of an offence.

KEYNOTE

This power applies to 'a vehicle' and is not limited to a motor, or even a mechanically propelled, vehicle.

'Neglecting' would appear to have a similar meaning to 'failing' (see e.g. *Pontin* v *Price* (1933) 97 JP 315).

Once the driver has stopped, he/she may commit this offence by moving off again in defiance of an order by the officer to stay put (see *Kentesber* v *Waumsley* [1980] Crim LR 383).

For a person to commit this offence, the constable must be engaged in the regulation of traffic.

If the constable was not so engaged and the direction given was to stop the vehicle, the appropriate charge would be under s. 163 of the Road Traffic Act 1988 (see below).

Unlike s. 163, however, there is no need for the constable to be in uniform but, if he/she is not, you would need to show that the defendant knew him/her to be a police officer.

Section 35 also applies to traffic wardens, **see para. 10.4.2**. Where a community support officer or accredited employee is exercising their authority in relation to the escorting of vehicles of exceptional dimensions (**see chapter 13**), this section will also apply to any directions properly given by them.

Section 35(2) provides a police power to direct traffic into a census area for the purpose of a traffic survey. That power must not be used in such a way as to cause unnecessary delay to someone who indicates that they are unwilling to provide information for the purposes of the survey (s. 35(3)).

10.4.2 Pedestrians Contravening Directions

OFFENCE: **Pedestrians Failing to Comply with Directions of Constable—*Road Traffic Act 1988, s. 37***

- Triable summarily • Fine

(No specific power of arrest)

The Road Traffic Act 1988, s. 37 states:

Where a constable in uniform is for the time being engaged in the regulation of vehicular traffic in a road, a person on foot who proceeds across or along the carriageway in contravention of a direction to stop given by the constable in the execution of his duty, either to persons on foot or to persons on foot and other traffic, is guilty of an offence.

KEYNOTE

The former restriction preventing traffic wardens from directing traffic from a moving vehicle has been removed (see the Functions of Traffic Wardens (Amendment) Order 2002 (SI 2002/2975, reg. 3(3)). Traffic wardens may also be used in the stopping of vehicles for roadside tests (as to which, **see para. 10.4.1** and **chapter 9**) and for escorting vehicles or trailers carrying loads of exceptional dimensions (**see chapter 13**).

OFFENCE: **Failing to Provide Name and Address—*Road Traffic Act 1988, s. 169***

- Triable summarily • Fine

(No specific power of arrest)

The Road Traffic Act 1988, s. 169 states:

> A constable may require a person committing an offence under section 37 of this Act to give his name and address, and if that person fails to do so he is guilty of an offence.

KEYNOTE

The requirement here is for the person to state his/her own name and address. However, given the decision in *R* v *McCarthy* (1999) *The Times*, 8 January, under a similar provision relating to the duty following an accident (**see para. 4.2**), it is uncertain whether the furnishing of some other address such as that of the defendant's solicitor would be enough to meet the requirement.

More generally, the police have specific powers to stop pedestrians in relation to authorised stop and search operations under certain conditions provided for by the Terrorism Act 2000 (**see General Police Duties, chapter 2**).

Traffic Wardens

By virtue of the Functions of Traffic Wardens Order 1970 (SI 1970/1958) and s. 96 of the Road Traffic Regulation Act 1984, ss. 35 and 37 above also apply to directions given by traffic wardens. Traffic wardens can also demand the name and address of any person committing an offence under s. 37.

This offence deals with pedestrians who, when directed to stop by a constable in uniform engaged in directing traffic, continue along the relevant road or across it.

Where a community support officer or accredited employee is exercising their authority in relation to the escorting of vehicles of exceptional dimensions (**see chapter 13**), this section will also apply to any directions properly given by them.

11 Driver licensing

11.1 Introduction

The law regulating driver licensing was greatly simplified by the introduction of the Motor Vehicles (Driving Licences) Regulations 1999 (SI 1999/2864). As with most statutory instruments in the area of road traffic, these principal Regulations have themselves been amended many times and it is important to refer to the most up to date list of such amendments when dealing with licensing matters. In cases of any doubt, the DVLA are generally the best source of information.

Most of the relevant legislation governing the licensing of drivers to drive motor vehicles can be found in the 1999 Regulations, together with:

- Part III of the Road Traffic Act 1988
- the Road Traffic (Driver Licensing and Information Systems) Act 1989
- the Road Traffic (New Drivers) Act 1995.

The 1999 Regulations, which have consolidated many of their predecessors, are complex and their exact wording should be consulted when considering offences or entitlements to drive. Infringement is generally charged under the Road Traffic Offenders Act 1988, s. 91.

Depriving a person of his/her driving licence can be a significant step in relation to his/her human rights generally. Where a person needs the licence for his/her livelihood, this penalty becomes even more significant (a fact which is taken into account by the limits placed on a court when imposing a disqualification order in child support cases). In another context, the Court of Appeal recognised the possibility that economic interests arising from the granting of a licence by a public authority might be treated as rights of property (e.g. for the purposes of Article 1, Protocol 1 of the European Convention) (*Kirk* v *Legal & General Assurance Ltd* [2002] IRLR 124). This approach finds further support in other cases, such as the decision of the Court of Appeal that, in deciding the fitness of goods vehicle operators, the Traffic Commissioners are bound by the general principles of proportionality under the European Convention and must consider the effect of their decision on the person's Convention rights (*Crompton* (*T/A David Crompton Haulage*) v *Department of Transport North Western Area* (2003) LTL 31 January).

11.2 Driving Otherwise than in Accordance with Licence

Since the implementation of the EC directive on driving licences (Council Directive 91/439/EEC) the categories of vehicles for the purposes of driving licences has been altered. The current categories are to be found in sch. 2 to the 1999 Regulations (**see appendix 6**).

OFFENCE: **Driving otherwise than in accordance with licence—** ***Road Traffic Act 1988, s. 87***

- Triable summarily • Fine • Discretionary disqualification for s. 87(1) under specified circumstances

(No specific power of arrest)

The Road Traffic Act 1988, s. 87 states:

(1) It is an offence for a person to drive on a road a motor vehicle of any class otherwise than in accordance with a licence authorising him to drive a motor vehicle of that class.

(2) It is an offence for a person to cause or permit another person to drive on a road a motor vehicle of any class otherwise than in accordance with a licence authorising that other person to drive a motor vehicle of that class.

KEYNOTE

This offence will apply where a person has a particular licence and has failed to abide by any conditions attached to it, or where a person does not hold a licence at all. In proving such an offence, you must show that the defendant drove a motor vehicle on a road (for the definitions of which, **see chapter 1**); it is then for the defendant to show that he/she had a licence to do so (see *John* v *Humphreys* [1955] 1 All ER 793).

Section 88 provides a long list of exceptions to the general prohibition imposed by s. 87. Those exemptions include:

- Drivers who have had their licence revoked on disqualification and who have re-applied for another.
- Drivers who have a 'qualifying application' lodged with the Secretary of State.
- Drivers from overseas becoming resident in Great Britain.

For the elements required in 'causing' and 'permitting', **see chapter 1**. Section 88 sets out a number of exemptions to the requirements of s. 87.

For overseas drivers, **see para. 11.6.3**.

For offences in relation to the supervision of learner drivers, **see para. 11.6.1**.

11.3 The Licence

A driving licence must be issued in the form prescribed by the Secretary of State under s. 98 of the Road Traffic Act 1988. From July 1998 a new photocard driving licence was introduced as part of the harmonisation process between our domestic road traffic legislation and that of the rest of the European Union (see the Driving Licences (Community Driving Licence) Regulations 1998 (SI 1998/1420)). Although the existing paper licences are still valid, the intention is to phase them out over time and current holders of such licences may exchange them for a new photocard licence. In addition to the photocard licence, there is a paper counterpart containing much of the same detail as the plastic card but not the holder's photograph.

Photocard licences are *pink* for full licence holders and *green* for provisional licence holders and need to be renewed every ten years until the holder is 70 years old. Licences held by drivers in Wales are printed in both Welsh and English.

Photocard licences are the same size as a credit card and contain the holder's:

- name
- address
- date of birth

- driver number
- driving entitlement
- photograph
- electronically-copied signature
- information codes showing any restrictions that apply to the holder.

The information codes, which appear in section 12 of the licence are reproduced in **appendix 10**. They can be of particular practical importance and will indicate, for instance, whether the driver is supposed to be wearing spectacles or some other form of visual correction. (For the power to require a driver to take an eyesight test, **see para. 11.6.4**.)

Licences will *generally* last until the holder's 70th birthday or for three years, whichever is the longer (see s. 99 of the Road Traffic Act 1988). On reaching 70, a driver may renew his/her licence every three years. This applies to full driving licences and most provisional licences. As noted above the photocard licence is renewable every 10 years.

Large goods vehicles and passenger carrying vehicles licences last until the drivers' 45th birthday or five years, whichever is the longer. If the driver is between 45 and 65 they last for five years or until the holder's 66th birthday, whichever is the *shorter*. After the driver has reached 65 the licence must be renewed annually (**see chapter 13**).

Provisional licences for motor bicycles of any class formerly lasted for two years. However, from 1 February 2001, new provisional licences for motor cycles remain generally valid until the holder's 70th birthday (reg. 15 as amended by the Motor Vehicles (Driving Licences) (Amendment) Regulations 2001 (SI 2001/53)). Full licences can act as provisional licences in many cases (except e.g. motor cycles over 125cc) and you should refer to the licence itself, together with the 1999 Regulations and the provisions of s. 97 in establishing whether a particular licence does entitle the holder to drive other vehicles as a provisional licence holder.

A person granted a driving licence must sign it in ink forthwith; failing to do so is an offence under reg. 20 and the Road Traffic Offenders Act 1988. Regulation 20 also makes provision for the signing of a document to allow the electronic reproduction of the driver's signature on the photocard licence (**see para. 11.3.2**).

A licence holder must surrender his/her licence to the Secretary of State on changing name or address and failure to do so is a summary offence under s. 99(5) of the Road Traffic Act 1988. The Secretary of State may also revoke driving licences under certain conditions (see e.g. s. 93 of the 1988 Act).

11.3.1 Police Powers

Power to Demand Production of Driving Licence

Section 164 of the Road Traffic Act 1988 states:

(1) Any of the following persons—
 (a) a person driving a motor vehicle on a road,
 (b) a person whom a constable or vehicle examiner has reasonable cause to believe to have been the driver of a motor vehicle at a time when an accident occurred owing to its presence on a road,
 (c) a person whom a constable or vehicle examiner has reasonable cause to believe to have committed an offence in relation to the use of a motor vehicle on a road, or
 (d) a person—
  (i) who supervises the holder of a provisional licence while the holder is driving a motor vehicle on a road, or
  (ii) whom a constable or vehicle examiner has reasonable cause to believe was supervising the holder of a provisional licence while driving, at a time when an accident occurred

> owing to the presence of the vehicle on a road or at a time when an offence is suspected of having been committed by the holder of the provisional licence in relation to the use of the vehicle on a road,
>
> must, on being so required by a constable or vehicle examiner, produce his licence and its counterpart for examination, so as to enable the constable or vehicle examiner to ascertain the name and address of the holder of the licence, the date of issue, and the authority by which they were issued.

KEYNOTE

In order to have met with the requirement to 'produce' a licence—or other documentation—the person must allow the constable (or vehicle examiner) a reasonable time to check the name and address of the holder, the date of issue and the authority under which the licence was issued (*Tremelling* v *Martin* [1971] RTR 196). Therefore simply flashing the licence at a police officer or waving it under his/her nose would not discharge the requirements of s. 164.

The requirement under s. 164(1)(a) is worded in the present tense. Therefore the power to require a licence under this subsection ceases when the driver ceases 'driving' (as to which, **see chapter 1**) (*Boyce* v *Absalom* [1974] RTR 248). Unfortunately the legislators did not use the same wording when drafting s. 164(1)(d)(i) in relation to supervisors of learner drivers. It is submitted, however, that the same restrictions would apply and that the power under this section would only apply if the relevant person were still supervising a learner at the time of the demand.

Clearly if a driver is reasonably believed to *have been* driving under the circumstances described in s. 164(1)(b), (c) and (d)(i) then the power to demand the relevant licence would not be restricted to the present tense.

Note that the requirement is to produce both the driving licence *and* its counterpart (**see para. 11.3**). It also applies to community licences (s. 164(11)).

A vehicle examiner is an examiner appointed under s. 66A of the 1988 Act (s. 164(11)).

Unlike some other driving documents (e.g. an insurance certificate, **see chapter 6**) driving licences must be produced in person (s. 164(8)).

Failing to produce a licence and its counterpart when lawfully required is a summary offence under s. 164(6).

Power to Demand Date of Birth

Section 164(2) of the Road Traffic Act 1988 states:

> (2) A person required by a constable under subsection (1) above to produce his licence must in prescribed circumstances, on being so required by the constable, state his date of birth.

KEYNOTE

The 'prescribed circumstances' are set out under reg. 83 of the Motor Vehicles (Driving Licences) Regulations 1999 and are:

- where the person *fails* to produce the licence *forthwith* (although the regulation says nothing about the *counterpart*); *or*
- where the person *produces* a licence which the police officer has reason to suspect
  - was not granted to that person
  - was granted to that person in error
  - contains an alteration to the particulars on the licence *other than the driver number*, made with intent to deceive; or
  - in which the driver number has been altered, removed or defaced; *or*
- where the person is/was a supervisor of a learner driver as specified under s. 164(1)(d) *and* the police officer has reason to suspect that he/she is under 21 years of age.

Making an alteration with intent to deceive may also amount to a more serious offence of deception/forgery (**see Crime, chapter 13**) or falsification of documents (**see chapter 16**). There does not need to be an 'intent to deceive'; simply reason for the officer to suspect such a motive.

Failing to state one's date of birth when lawfully required is a summary offence under s. 164(6).

Where a person does state his/her date of birth as required, the Secretary of State may serve a written notice on that person requiring him/her to provide evidence verifying that date of birth. If the person's name differs from his/her name at birth, the Secretary of State may also serve a similar notice requiring a statement from the person as to his/her name when born (s. 164(9)). Knowingly failing to do so is a summary offence (s. 164(9)).

Other Powers

Where a police officer has reasonable cause to believe that the holder of a driving licence has knowingly made a false statement for obtaining it, the officer may require the production of that licence and its counterpart (s. 164(4)). Failing to do so is a summary offence under s. 164(6).

Where the rider of a motor bicycle (**see chapter 1**) produces a provisional driving licence to a police officer and that officer has reasonable cause to believe that the holder was not driving the motor bicycle as part of the training on an approved course, he/she may require the production of a prescribed certificate in relation to the completion of, or exemption from such a course (s. 164(4A)). Failing to produce such a certificate when lawfully required is a summary offence under s. 164(6).

Licence Revoked

Where a person has had his/her licence revoked in relation to a disability or because it has expired and he/she fails to deliver the licence to the DVLA, a police officer or vehicle examiner may require the licence to be produced and may then seize it (s. 164(3)). Similar powers apply where the licence holder has been ordered to produce his/her licence to a court (s. 164(5)). Failing to do so is a summary offence under s. 164(6).

Licences Surrendered under Fixed Penalty Procedure

If a person has surrendered his/her licence and its counterpart to a police officer in connection with a fixed penalty offence (**see chapter 14**) he/she cannot produce it if required under s. 164. Therefore s. 164(7) provides that, if such a person produces:

- a *current* receipt (under s. 56 of the Road Traffic Offenders Act 1988) to that effect (either then and there or within seven days) and
- *if required to do so*, produces the licence and counterpart on his/her return to a police station

he/she does not commit an offence under s. 164(6).

Defence

Section 164(8) states:

> (8) In proceedings against any person for the offence of failing to produce a licence and its counterpart it shall be a defence for him to show that—
> (a) within seven days after the production of his licence and its counterpart was required he produced them in person at a police station that was specified by him at the time their production was required, or
> (b) he produced them in person there as soon as was reasonably practicable, or
> (c) it was not reasonably practicable for him to produce them there before the day on which the proceedings were commenced,
>
> and for the purposes of this subsection the laying of the information . . . shall be treated as the commencement of the proceedings.

KEYNOTE

This general defence allows for the issuing of an HORT/1 in respect of driving licences and counterparts; it also applies to the certificate showing that a licence holder has completed his/her compulsory training in relation to motor bicycles or that he/she is exempt from having to complete such a course (s. 164(8A)). Section 164(8) and (8A) also allow for occasions where the production of the relevant documents is not possible at the time.

11.3.2 Offence

It is a summary offence—under reg. 20 of the Motor Vehicles (Driving Licences) Regulations 1999 and s. 91 of the Road Traffic Offenders Act 1988—to fail to sign a driving licence immediately it is received by the holder. Holders of the photocard licence (**see para. 11.3**) are required to provide a signature on the relevant form in order that the DVLA can reproduce his/her signature electronically on that photocard and reg. 20 has been amended accordingly.

11.3.3 Categories and Ages

Section 87 of the Road Traffic Act 1988 requires that any person driving a motor vehicle on a road does so in accordance with a licence authorising him/her to drive a motor vehicle *of that class* (**see para. 11.2 and appendix 6**).

The classification of vehicles is created by virtue of reg. 4 of the Motor Vehicles (Driving Licences) Regulations 1999 (SI 1999/2864) and sch. 2 to those Regulations. Schedule 2 is set out at **appendix 6**. In relation to licences granted before 1 January 1997, provisions for the change from the 'old' categories to the new ones are to be found in reg. 76.

Regulation 7 of the 1999 Regulations sets out the categories of vehicle that a driver will be deemed as competent to drive by virtue of holding a licence. This regulation restricts drivers who passed tests in relation to certain restricted classes of vehicle (e.g. those with an automatic transmission) to vehicles falling within that category or sub-category. Similar provisions apply under reg. 7 to drivers having adaptations in relation to a disability. The categories of vehicle that a licence holder is entitled to drive will be clearly marked on his/her licence and its counterpart and any relevant restrictions will appear in the information codes in section 12 (**see appendix 10**).

The ages at which a person may drive the relevant category are generally to be found in the Road Traffic Act 1988, s. 101—although, confusingly, some of the restrictions on additional categories under reg. 7 above also include references to minimum ages. The table under s. 101 is set out at **appendix 1**.

That table is amended in certain cases by the provisions of reg. 9 of the 1999 Regulations. Those amendments are principally in the areas of:

- Large motor bicycles (**see chapter 1**) where, under certain circumstances, the age is raised to 21 years (unless the person has passed a relevant test or the vehicle is owned by the Secretary of State for Defence/is being used for military purposes by a member of the armed forces).
- Some small agricultural and forestry tractors without trailers or with smaller trailers, where the limit is lowered to 16 years.
- Small vehicles (**see chapter 1**) without trailers driven by someone in receipt of certain disability allowances where the limit is lowered to 16 years.
- Medium-sized goods vehicles (**see chapter 1**) drawing a trailer where the maximum authorised mass of the combination exceeds 7.5 tonnes where the limit is raised to 21 years.

- Large goods and large passenger carrying vehicles (**see chapter 1**) driven under certain circumstances in relation to NHS and Primary Care Trust ambulances, training scheme employees and holders of provisional or full PSV licences operating under a PSV operators' licence (**see chapter 13**). In the specified circumstances, the limit is lowered to 18 years.
- Some vehicles owned or operated by the armed forces and some road rollers.

These amendments are very detailed and are frequently updated. Therefore reference should be made to the Regulations themselves wherever possible and again the DVLA are probably the best source of up-to-date information in this respect.

11.4 Driving Tests

Sections 89 and 89A of the Road Traffic Act 1988 impose the need to pass a prescribed driving test before being granted a licence, while part III of the Motor Vehicles (Driving Licences) Regulations 1999 (as amended) makes provisions for those driving tests—along with the various exemptions from any part of the test that may apply. The content of the test itself was recently changed and now includes a two-part theory test, with the second part covering hazard perception.

The 1999 Regulations set out the requirements for the testing of drivers for specific categories of vehicle and introduce a two-part test (similar to that taken by car drivers) for drivers wishing to drive medium and large goods/passenger vehicles. Schedules 7 and 8 contain the syllabus for those tests. Schedule 12 contains elements for approved training courses for motor cycles.

Schedule 6 requires driving test candidates to produce proof of identity before taking their test. Under regs 23 and 24, chief officers of police may conduct driving tests for their personnel under certain conditions.

A new section (s. 99ZA) was inserted into the Road Traffic Act 1988 from 1 May 2002. This effectively allows the Secretary of State to make regulations requiring people to complete driving training courses before taking certain driving tests or before driving certain vehicles on a provisional licence.

11.5 Disqualified Drivers

Most disqualifications are imposed by the courts under part II of the Road Traffic Offenders Act 1988. Some special provisions exist in relation to the disqualification of holders of large goods or passenger-carrying vehicles (see the Road Traffic Act 1988 s. 115 and the Driving Licences (Community Driving Licence) Regulations 1998 (SI 1998/1420)). Many drivers also find themselves disqualified by reason of the 'totting up' procedure whereby, once you have accumulated 12 penalty points within three years of the commission of the first offence, you are automatically disqualified. There is a separate, more stringent, system in place for new drivers (**see para. 11.8**). In addition, s. 146 of the Powers of Criminal Courts (Sentencing) Act 2000 makes provision for courts to disqualify defendants convicted of any offence, instead of or as well as any other punishment. Section 147 allows the Crown Court to impose disqualification where the person has been convicted of an offence punishable on indictment with two years or more imprisonment, or any offence involving assault (as to which, **see Crime, chapter 8**).

Courts also have powers to impose disqualification on drivers under other statutes. Under the Crime (Sentences) Act 1997, any court which has convicted a person of any offence may generally order the person to be disqualified from holding or obtaining a licence in addition to, or instead of dealing with the person in any other way. This disqualification may be for such period as the court thinks fit.

In relation to fine defaulters, a *magistrates' court* may order that a person be disqualified from holding or obtaining a licence *instead* of proceeding against the person under its other statutory powers for dealing with defaulters. This order may be imposed for a *maximum of 12 months*.

Finally, where it has been shown before a *magistrates' court* that a person has demonstrated wilful refusal or culpable neglect in relation to their responsibilities under the Child Support Act 1991, that court may order the person to be disqualified from holding or obtaining a licence (s. 40). Any such order may be imposed for a *maximum of two years* and may be suspended for such time and carry such conditions as the court thinks just (see the Child Support, Pensions and Social Security Act 2000, s. 16). There are additional conditions that must be met before a court can impose a disqualification order on a person under this provision (see s. 39).

At the time of writing, the Government has published a consultation document towards ratifying the EU Driving Disqualification Convention The aim of the consultation is to extend the effects of disqualification in one Member State across all others, thus preventing drivers who are disqualified in one country from continuing to drive in another.

OFFENCE: **Driving while Disqualified—*Road Traffic Act 1988, s. 103(1)(b)***

- Triable summarily • Six months' imprisonment and/or a fine
- Discretionary disqualification

(Arrestable offence)

The Road Traffic Act 1988, s. 103 states:

(1) A person is guilty of an offence if, while disqualified for holding or obtaining a licence, he—
 (a) ...
 (b) drives a motor vehicle on a road.

OFFENCE: **Obtaining Licence while Disqualified—*Road Traffic Act 1988, s. 103(1)(a)***

- Triable summarily • Fine

(No specific power of arrest)

The Road Traffic Act 1988, s. 103 states:

(1) A person is guilty of an offence if, while disqualified for holding or obtaining a licence, he—
 (a) obtains a licence...

KEYNOTE

A licence obtained by a person who is disqualified is of no effect (s. 103(2)).

The Court of Appeal has recently taken the opportunity to set out the key issues to be considered when looking at an offence of driving while disqualified. In that case (*Shackleton* v *Chief Constable of Lancashire Police* (2001) LTL 30 October), the defendant was known to be a disqualified driver and was seen by a police officer to be 'jogging' away from a parked Ford Escort. The officer followed and arrested him under the former statutory power. In hearing the defendant's appeal against the lawfulness of his arrest, the Court said that the primary issue was whether the defendant had driven the car or whether the police officer had the reasonable belief that he had driven it. Guidance on this issue was to be found in *Pinner* v *Everett* [1969] 1 WLR 1266 where the House of Lords concluded that each case had to be considered on its merits. Their Lordships had also held that there was no requirement for the vehicle to be in motion and that the key considerations in assessing each case on its merits were whether the defendant:

- had actually stopped driving or intended to carry on driving (e.g. at a set of traffic lights)
- was still driving
- had arrived at his/her destination or intended to continue to a further location
- had been prevented or persuaded from driving by someone else.

If the person is disqualified *by reason of age* (under s. 101), neither this offence *nor the power of arrest* will apply; the relevant offence would be under s. 87(1) (**see para. 11.2**).

The offence of driving while disqualified is one of strict liability (**see Crime, chapter 1**) and there is no need to show that the driver knew of the disqualification (*Taylor* v *Kenyon* [1952] 2 All ER 726).

You must *prove* that the person who was driving the motor vehicle was in fact disqualified. This may sound pretty obvious but it has occasionally been overlooked causing problems at trial (*R* v *Derwentside Magistrates' Court, ex parte Heaviside* [1996] RTR 384). Although one—and perhaps the best—way of proving that a defendant was disqualified at the time of driving is by a certificate under s. 73(4) of the Police and Criminal Evidence Act 1984, this is by no means the *only* way of so proving. The courts have accepted the evidence of someone who was present in court at the time the person was disqualified (*Derwentside* above); they have also accepted the defendant's own admission that he/she was disqualified, both in interview under caution and in evidence before the court (see *Moran* v *CPS* (2000) 164 JP 562 and *DPP* v *Mooney* [1997] RTR 434).

A person may still commit the offence(s) above even if the disqualification is later quashed on appeal; the offence is complete as long as he/she was disqualified at the relevant time (*R* v *Lynn* [1971] RTR 369).

The elements of 'driving', 'motor vehicle' and 'road' are set out in **chapter 1**.

As the offence of driving while disqualified is a summary offence, it cannot be 'attempted' (**see Crime, chapter 3**). Therefore, although this offence has become an 'arrestable offence', this change in legislation has done little to improve the preventive effect of police powers. For a full discussion of powers of arrest and their limitations here, **see General Police Duties, chapter 2**.

A similar practical problem arises under the offence of taking a conveyance without the owner's consent (**see Crime, chapter 12**). Strangely the endorsement offence codes (**see appendix 11**) and sch. 2 to the Road Traffic Offenders Act 1988 (**see appendix 3**) respectively make reference to 'attempting' these offences.

If a passenger genuinely believes that the driver is entitled to drive the vehicle in which they are both found, that belief will prevent him/her being charged with 'aiding and abetting' the offence under s. 103(1)(b) (*Bateman* v *Evans* (1964) 108 SJ 522). In the case of someone supervising a 'learner driver' who turns out to be disqualified, it seems reasonable to expect a supervisor to establish that the driver under his/her supervision has a current and valid licence before venturing out onto the road.

The general and growing defence of 'duress of circumstances' (**see chapter 2** and **Crime, chapter 4**) applies to the offence under s. 103(b).

11.5.1 Disqualification until Test is Passed

Section 36 of the Road Traffic Offenders Act 1988 states:

(1) Where this subsection applies to a person the court must order him to be disqualified until he passes the appropriate driving test.

(2) Subsection (1) above applies to a person who is disqualified under section 34 of this Act on conviction of—
   (a) manslaughter, or in Scotland culpable homicide, by the driver of a motor vehicle, or
   (b) an offence under section 1 (causing death by dangerous driving) or section 2 (dangerous driving) of the Road Traffic Act 1988.

(3) Subsection (1) above also applies—
   (a) to a person who is disqualified under section 34 or 35 of this Act in such circumstances or for such period as the Secretary of State may by order prescribe, or

(b) to such other persons convicted of such offences involving obligatory endorsement as may be so prescribed.

(4) Where a person to whom subsection (1) above does not apply is convicted of an offence involving obligatory endorsement, the court may order him to be disqualified until he passes the appropriate driving test (whether or not he has previously passed any test).

(5) In this section—

'appropriate driving test' means—

(a) an extended driving test, where a person is convicted of an offence involving obligatory disqualification or is disqualified under section 35 of this Act,

(b) a test of competence to drive, other than an extended driving test, in any other case,

'extended test' means a test of competence to drive prescribed for the purposes of this section, and 'test of competence to drive' means a test prescribed by virtue of section 89(3) of the Road Traffic Act 1988.

KEYNOTE

Section 36 gives the court the power to disqualify a person from holding or obtaining a licence (s. 98(1)) until the person passes an 'appropriate' test. If the person is disqualified under the provisions for 'totting up' penalty points (see s. 35) or is found guilty of an offence involving obligatory disqualification (see s. 34), the 'appropriate' test is an *extended* test as defined at s. 36(5)(b). In other cases the 'appropriate' test will be a regular driving test under s. 89(3) of the Road Traffic Act 1988.

The relevant offences are set out at s. 36(2)(a) and (b), namely manslaughter, causing death by dangerous driving and dangerous driving itself.

Section 36(3)(b) enables the Secretary of State to add further offences to the list. This has been done in relation to an offence under s. 3A of the Road Traffic Act 1988—causing death by careless driving when under the influence of drink or drugs (see the Driving Licences (Disqualification until Test Passed) (Prescribed Offence) Order 2001 (SI 2001 No. 4051)). In the case of an offence under s. 3A the relevant test is an extended driving test.

Where a person is convicted of an offence involving obligatory disqualification that is not covered by s. 36(1) and (2) (e.g. an offence under s. 4(1) of the Road Traffic Act 1988; **see para. 5.2**), the court *may* order disqualification until that person passes an appropriate driving test (s. 36(4)). Here the appropriate test will generally be the regular driving test (unless the disqualification involves 'totting up'). In determining whether or not to disqualify a person in such a case, the court must have regard to the safety of road users (s. 36(6)).

Section 36(7) to (13) make provisions for the extent and effect of a disqualification until a test is passed and also provides for tests of competence in certain other European countries to be treated as meeting the requirements of s. 36.

A person who has been disqualified until he/she has passed a test can apply for a provisional licence and can drive in accordance with the conditions of such a licence (s. 37(3)). After all, it is the purpose of such a disqualification that the offender be required to prove their competence to drive, rather than simply removing them from the road.

Where a person so disqualified obtains a provisional licence and then drives without supervision or 'L' plates, he/she commits the offence of driving while disqualified under s. 103 of the Road Traffic Act 1988 above (*Scott* v *Jelf* [1974] RTR 256).

The courts have a power to impose an interim disqualification under certain conditions. Where a defendant has been convicted of a relevant offence under ss. 34–36 and the magistrates' court commits him/her to the Crown Court *for sentence*, the court may order an interim disqualification (see s. 26(1) of the Road Traffic Offenders Act 1988 and s. 6 of the Powers of Criminal Court (Sentencing) Act 2000). The magistrates' court may also impose an interim disqualification when:

- remitting the person to another magistrates' court (under s. 10 of the Powers of Criminal Court (Sentencing) Act 2000);
- deferring passing a sentence on the person; or
- adjourning after convicting the person but before dealing with him/her for the offence.

A magistrates' court cannot impose an interim disqualification when committing a defendant *for trial*.

11.5.2 After Disqualification has Expired

- Once a period of disqualification ends, the person may apply for another licence.
- On application for another licence the person falls within the category of someone who 'has held and is entitled to obtain' a licence under s. 88 of the Road Traffic Act 1988.
- Section 88 provides an exemption to the offence of driving otherwise than in accordance with a licence (s. 87) (**see para. 11.3.3**).
- The person can begin to drive again as soon as a proper application has been received by the Driver and Vehicle Licensing Agency.

If a person drives before applying for a new licence he/she commits the offence under s. 87 (**see para 11.3.3**).

A person disqualified until he/she passes a driving test may (or arguably *must*) apply for a provisional licence. On application he/she may begin to drive in accordance with the conditions below.

11.6 Learner Drivers

Learner drivers are generally required to hold provisional licences (**see para. 11.3**). The granting of provisional licences is governed by s. 97(3) of the Road Traffic Act 1988 which states:

> (3) A provisional licence—
> (a) shall be granted subject to prescribed conditions,
> (b) shall, in any cases prescribed for the purposes of this paragraph, be restricted so as to authorise only the driving of vehicles of the classes so prescribed,
> (c) may, in the case of a person appearing to the Secretary of State to be suffering from a relevant disability or a prospective disability, be restricted so as to authorise only the driving of vehicles of a particular construction, or design specified in the licence, and . . .

KEYNOTE

Drivers—or potential drivers—may apply for provisional licences specifically to drive certain categories of vehicles or they may be able to use a licence that they already hold. The type of provisional entitlement that each category of vehicle requires is governed by s. 97 of the Road Traffic Act 1988 and reg. 11 of the Motor Vehicles (Driving Licences) Regulations 1999.

Some sub-categories of vehicle can only be driven by certain learner drivers who hold a full licence in another, specified category. These categories are set out in the table found under reg. 11 (**see appendix 6**). Regulation 11 makes exemptions in relation to full-time members of the armed forces.

Further provision is made (by s. 98(2) of the Road Traffic Act 1988) to allow full licences in some categories to count as provisional licences for others. These categories are set out in the table found under reg. 19 (**see appendix 6**).

Motor Bicycles and Mopeds

Section 97 of the Road Traffic Act 1988 goes on to state that a provisional licence:

> (3) . . .
> (d) shall not authorise a person under the age of 21 years, before he has passed a test of competence to drive a motor bicycle,—
> (i) to drive a motor bicycle without a side-car unless it is a learner motor bicycle (as defined in subsection (5) below) or its first use (as defined in regulations) occurred before January 1, 1982 and the cylinder capacity of its engine does not exceed 125 cubic centimetres, or
> (ii) to drive a motor bicycle with a side-car unless its power to weight ratio is less than or equal to 0.16 kilowatts per kilogram.

(e) except as provided under subsection (3B) below, shall not authorise a person, before he has passed a test of competence to drive, to drive on a road a motor bicycle or moped except where he has successfully completed an approved training course for motor cyclists or is undergoing training on such a course and is driving the motor bicycle or moped on the road as part of the training.

(3A) Regulations may make provision as respects the training in the driving of motor bicycles and mopeds of persons wishing to obtain licences authorising the driving of such motor bicycles and mopeds by means of courses of training provided in accordance with the regulations; and the regulations may in particular make provision with respect to—

(a) the nature of the courses of training;

(b) the approval by the Secretary of State of the persons providing the courses and the withdrawal of his approval;

(c) the maximum amount of any charges payable by persons undergoing the training;

(d) certificates evidencing the successful completion by persons of a course of training and the supply by the Secretary of State of the forms which are to be used for such certificates; and

(e) the making, in connection with the supply of forms of certificates, of reasonable charges for the discharge of the functions of the Secretary of State under the regulations; and different provision may be made for training in different classes of motor bicycles and mopeds.

KEYNOTE

The restrictions at s. 97(3)(d)(i) and (ii) apply to motor bicycles—which includes a sidecar combination. Therefore learner riders are generally restricted to using 'learner motor bicycles', motor *cycles* (i.e. mainly those having more than two wheels) and older motor bicycles first used before 1 January 1982 whose engine capacity does not exceed 125 cc. For the respective definitions, **see chapter 1**.

Unless a driver is exempted by the Regulations, he/she may not drive a motor bicycle or moped (**see chapter 1**) on a road before passing an approved training course or while on such a course as part of that training (s. 97(3)(e)). The regulation of approved training courses (also known as compulsory basic training—'CBT') for riders of motor bicycles is governed by part V of the 1999 Regulations. Among the other qualifications that such a person must have, an approved CBT instructor must show the Secretary of State that he/she is a 'fit and proper person'. (For the situation regarding the provision of driver instruction generally, **see para. 11.6.2**.)

On completion of a CBT course a person will be issued with a certificate. After 1 February 2001 this certificate will be valid for two years (reg. 68 as amended by the Motor Vehicles (Driving Licences) (Amendment) Regulations 2001 (SI 2001 No. 53)).

As with other provisional licence holders, learner riders will be subject to the conditions set out in reg. 16 (see below). The requirement to have a qualified driver supervising him/her on the vehicle does not apply when the licence holder is riding a moped, or motor bicycle with or without a sidecar (reg. 16(3)(b)).

Regulation 16(6) provides that the holder of a provisional licence authorising the driving of a moped, or motor bicycle with or without a sidecar, shall not drive such a vehicle while carrying another person. This creates an absolute ban on passengers, qualified or otherwise, on such vehicles being driven by provisional licence holders.

A person learning to ride a 'large motor bicycle' must:

- Hold a provisional licence authorising the driving of motor bicycles other than learner motor bicycles (category A1).
- Be at least 21 years old (see below) and must abide by the requirements of reg. 16(7).

Regulation 16(7) requires such learners to:

- be in the presence and under the supervision by a certified direct access instructor
- be able to communicate with the instructor by means of a non hand-held radio system
- be wearing apparel (together with the instructor) which is fluorescent or, if during the hours of darkness, is either fluorescent or luminous.

An exception to the requirement for the learner to be in radio contact with the instructor has been inserted to cover the situation where the learner has a hearing impairment (reg. 16(11)). In such cases, it will suffice that the learner and the instructor employ a 'satisfactory means of communication' which they agree upon *before the start of the journey*.

The requirements that must be met by a 'direct access instructor' are set out under reg. 65. Note that, while conducting the above training, the maximum number of 'trainees' that a direct access instructor can supervise is two (reg. 67).

As a further result of the Motor Vehicles (Driving Licences) (Amendment) Regulations 2001 above, additional requirements have been imposed on holders of provisional licences to drive learner motor bicycles or mopeds. These regulations add reg. 16(7A) to the 1999 Regulations. In essence, reg. 16(7A) says that the holder of a provisional licence for a learner motor bicycle or moped must not drive such a vehicle on a road when undergoing training (other than as part of an approved motor cyclists' training course) by a paid instructor unless the instructor is at all times present with him/her on the road and is supervising no more than three other such provisional licence holders.

Full Licences used as Provisional

As mentioned above, s. 98(2) of the Road Traffic Act 1988 provides that a person holding a full licence for certain classes of vehicle may drive motor vehicles of other classes as if authorised by a provisional licence for those other classes. However, reg. 19(5) of the 1999 Regulations modifies the effect of s. 98(2) in relation to motor bicycles. In the case of a licence that authorises the driving only of learner or standard motor bicycles, s. 98(2) will not apply so as to authorise the driving of a large motor bicycle by a person under the age of 21. Therefore a person who holds a licence for a learner or standard motor bicycle but who is under the age of 21 cannot use that licence as a provisional licence for a large motor bicycle. Further restrictions are imposed on holders of full licences for 'standard' motor bicycles who wish to use the licence as a provisional licence to drive large motor bicycles.

General Requirements for Provisional Licence Holders

Holders of provisional licences must not drive or ride a motor vehicle of a class authorised by the licence unless he/she is under the supervision of a 'qualified driver' (see below) (reg. 16(2)(a) of the 1999 Regulations). He/she must also display the 'distinguishing mark' (L plate, **see appendix 7**) in the form set out at part 1 to sch. 4 in such a manner that it is clearly visible to other persons using the road from within a reasonable distance from the front and back of the vehicle (reg. 16(2)(b)). Provisional licence holders driving a motor vehicle on a road in Wales may display the alternative distinguishing mark (a 'D' plate, signifying *dysgwr* or 'learner' (reg. 16(4)).

The requirement to have a qualified driver supervising him/her on the vehicle does not apply to invalid carriages or vehicles in certain categories that are constructed or adapted to carry only one person (reg. 16(3)), neither does it apply to motor bicycles or mopeds (see above).

Provisional licence holders are generally prohibited from drawing trailers although certain classes of vehicles are exempt from this provision (reg. 16(2)(c)).

If a person is disqualified until he/she passes another test and then fails to abide by the provisions of the new provisional licence, he/she commits the offence of driving while disqualified (under s. 103 of the Road Traffic Act 1988), **see para. 11.5**.

If a person is a 'new driver' and has his/her licence revoked after accumulating six penalty points (under s. 3 of the Road Traffic (New Drivers) Act 1995, **see para. 11.8**), he/she will not be counted as 'disqualified' but will revert to the position of a learner driver.

If a provisional licence holder fails to observe these conditions, he/she commits an offence under s. 91 of the Road Traffic Offenders Act 1988 (i.e. breaching the Regulations);

if the person does *not have* a provisional licence he/she commits the offence under s. 87 of driving otherwise than in accordance with a licence (**see para. 11.3.3**).

Between Driving Test and Issue of Licence

Regulation 16(10) of the 1999 Regulations makes provision for the situation where a provisional licence holder has been issued the relevant certificate stating that he/she has passed a driving test but has not yet received his/her full licence. In such cases, the requirements to display 'L' (or 'D') plates and to be supervised, together with the restrictions on the carrying of passengers are removed.

11.6.1 Supervision of Learner Drivers

There is only one acceptable minimum standard of driving and all drivers, including learners, must observe it; if not, they commit the offence of careless driving (*McCrone* v *Riding* [1938] 1 All ER 157). (**See also chapter 2.**)

A supervisor of a learner driver is required, not to provide tuition to the learner, but to 'supervise'. That means doing whatever might reasonably be expected to prevent the learner driver from acting carelessly or endangering others (see *Rubie* v *Faulkner* [1940] 1 All ER 285). The duty includes being in a position to take control of the vehicle in an emergency. If the supervisor is not able to do this, either because of his/her physical state (e.g. drunk) or their being out of the vehicle (e.g. giving directions), the condition will not have been fulfilled.

Whether or not a supervisor fulfilled his/her duty will be a question of fact for a court to determine in each case. If it can be shown that the qualified person was not actually 'supervising' the driver then the requirement under reg. 16 above will not have been observed and the offence (under s. 91 of the Road Traffic Offenders Act 1988) will be committed. That offence will be aided and abetted (**see Crime, chapter 2**) by the so-called supervisor.

A 'supervisor' can also be convicted of aiding and abetting where the learner driver is over the prescribed limit or unfit through drink or drugs (*Crampton* v *Fish* (1969) 113 SJ 1003). (**See also chapter 5.**)

The duties of a supervisor extend to ensuring compliance with other legislative requirements made of drivers such as remaining at the scene of an accident (*Bentley* v *Mullen* [1986] RTR 7). (**See also chapter 4.**)

The supervisor must be a 'qualified driver' under reg. 17 of the 1999 Regulations which now states:

> (1) Subject to paragraph (2), a person is a qualified driver for the purposes of regulation 16 if he—
> (a) is 21 years of age or over,
> (b) holds a relevant licence,
> (c) has the relevant driving experience, and
> (d) in the case of a disabled driver, he is supervising a provisional licence holder who is driving a vehicle of a class included in category B and would in an emergency be able to take control of the steering and braking functions of the vehicle...

KEYNOTE

A 'relevant licence' means a full licence (including a Northern Ireland or Community licence) authorising the driving of vehicles of the same class as the vehicle being driven by the provisional licence holder (reg. 17(3)(c)). In the case of disabled drivers it means a full licence authorising the driving of a class of vehicles in category B other than invalid carriages. The reference to paragraph (2) is to people who are members of the armed forces acting in the course of their duties. 'Relevant driving experience' is defined at reg. 17(3)(d). *Generally* a person will have relevant driving experience if they have held the relevant full licence for a continuous or aggregate period of not less than three years. Other conditions setting out 'relevant driving experience' are imposed by reg. 17(3) in relation to the supervision of people driving vehicles in categories C, D, C + E and D + E.

OFFENCE: **Supervisor of Learner Driver Failing to Give Details—*Road Traffic Act 1988, s. 165(5)***

• Triable summarily • Fine

(No specific power of arrest)

The Road Traffic Act 1988, s. 165 states:

(5) A person—

(a) who supervises the holder of a provisional licence granted under Part III of this Act while the holder is driving on a road a motor vehicle (other than an invalid carriage), or

(b) whom a constable or vehicle examiner has reasonable cause to believe was supervising the holder of such a licence while driving, at a time when an accident occurred owing to the presence of the vehicle on a road or at a time when an offence is suspected of having been committed by the holder of the provisional licence in relation to the use of the vehicle on a road,

must, on being so required by a constable or vehicle examiner, give his name and address and the name and address of the owner of the vehicle.

KEYNOTE

Regulation 80 makes provision for holders of foreign licences to be treated as holders of relevant driving licences under part III of the Road Traffic Act 1988 for certain purposes. These requirements only apply in relation to the requirements of s. 87(1) (driving otherwise than in accordance with a licence (**see para. 11.3.3**)) and therefore do not entitle such licence holders to supervise learner drivers. Holders of such licences may, if they take out a provisional licence in the UK, not have to comply with the general requirements imposed on learner drivers (see above) (reg. 18).

11.6.2 Instruction of Learner Drivers

Anyone is able to give driving lessons provided they do not charge money or money's worth in return. If a person wants to give driving instruction for payment, he/she must be registered in accordance with the provisions of Part V of the Road Traffic Act 1988 (see s. 123). The purpose of the regulation of driving instructors was examined in *Mahmood* v *Vehicle Inspectorate* (1998) 18 WRTLB 1. There it was held by the Divisional Court that notions of contractual payment under civil law were not particularly helpful or relevant. What mattered was whether or not the defendant (instructor) had some sort of arrangement with the learner driver and that the arrangement had a 'commercial flavour'.

Police driving instructors are exempt from this requirement (s. 124).

Free driving lessons offered by someone in the business of buying and selling cars will be deemed to be given for payment if they are given in connection with the supply of a vehicle (s. 132(3)).

These restrictions only apply to 'motor cars' as specifically designed for this purpose (**see chapter 1**) therefore it would not prevent someone charging for driving lessons on motor cycles or other vehicles falling outside the definition (for motor bicycles, **see chapter 1**).

The relevant conditions for registered driving instructors can be found in the Motor Cars (Driving Instruction) Regulations 1996 (SI 1996/1983), as amended.

The relevant source for the content of test instruction requirements is the Motor Cars (Driving Instruction) Regulations 1989 (SI 1989/2057).

The Road Traffic (Driving Instruction by Disabled Persons) Act 1993 introduced provisions for driving lessons to be provided by disabled people and those provisions can be found in ss. 125A and 125B and ss. 133A–133D of the Road Traffic Act 1988.

11.6.3 Drivers from Other Countries

The entitlement of drivers living outside the United Kingdom to drive here under the authority of their overseas permits is governed by the Motor Vehicles (International

Circulation) Order 1975 (SI 1975/1208). If such drivers hold a domestic or Convention driving permit issued abroad or a British Forces driving licence, they may drive the vehicles covered by those authorities in Great Britain for one year (Article 2). Regulation 80 of the Motor Vehicles (Driving Licences) Regulations 1999 makes similar provisions in relation to people who become resident in the United Kingdom. That permit allows the holder to take a driving test in that 12 month period. If they do not do so successfully then they will need a GB provisional licence. The permit alone does not allow them to supervise learner drivers (**see para. 11.6.1**).

Visitors and new residents holding a valid driving licence may also use that licence during the first 12 months and, if they apply for a GB provisional licence during that period, they will be exempt from the conditions imposed on provisional licence holders (reg. 18).

Members of visiting forces and their dependants are covered by the Motor Vehicles (International Circulation) Order 1975, as amended (Article 3).

The former requirement for drivers from EU Member States to exchange their licences within one year of becoming a resident in Great Britain has been removed. The law governing such drivers can now be found in the Driving Licences (Community Driving Licence) Regulations 1996 (SI 1996/1974).

Drivers from Member States must meet the fitness requirements of British drivers (**see para. 11.6.4**) and, provided they are not disqualified, drivers meeting those physical requirements may exchange their licence for a part III licence if they have become normally resident in Great Britain (s. 89 of the Road Traffic Act 1988).

The 1988 Act also makes provision for the exchange of other licences from some non-EU countries.

11.6.4 Physical Fitness and Disability

Section 92 of the Road Traffic Act 1988 states:

(1) An application for the grant of a licence must include a declaration by the applicant, in such form as the Secretary of State may require, stating whether he is suffering or has at any time (or, if a period is prescribed for the purposes of this subsection, has during that period) suffered from any relevant disability or any prospective disability.

(2) In this Part of this Act—
'disability' includes disease and the persistent misuse of drugs or alcohol, whether or not such misuse amounts to dependency,
'relevant disability' in relation to any person means—
(a) any prescribed disability, and
(b) any other disability likely to cause the driving of a vehicle by him in pursuance of a licence to be a source of danger to the public, and
'prospective disability' in relation to any person means any other disability which—
(a) at the time of the application for the grant of a licence or, as the case may be, the material time for the purposes of the provision in which the expression is used, is not of such a kind that it is a relevant disability, but
(b) by virtue of the intermittent or progressive nature of the disability or otherwise, may become a relevant disability in course of time.

(3) If it appears from the applicant's declaration, or if on inquiry the Secretary of State is satisfied from other information, that the applicant is suffering from a relevant disability, the Secretary of State must, subject to the following provisions of this section, refuse to grant the licence.

KEYNOTE

Part VI of the Motor Vehicle (Driving Licences) Regulations 1999 sets out the specific requirements as to the physical fitness of drivers and include the standard of eyesight demanded of drivers of different vehicle groups; it also makes specific provisions for drivers with controlled epilepsy.

OFFENCE: **Driving with Uncorrected Defective Eyesight—*Road Traffic Act 1988, s. 96(1)***

- Triable summarily • Fine

(No specific power of arrest)

The Road Traffic Act 1988, s. 96 states:

(1) If a person drives a motor vehicle on a road while his eyesight is such (whether through a defect which cannot be or one which is not for the time being sufficiently corrected) that he cannot comply with any requirement as to eyesight prescribed under this Part of this Act for the purposes of tests of competence to drive, he is guilty of an offence.

KEYNOTE

Under s. 96(2) a constable having reason to suspect that a person driving a motor vehicle may be guilty of this offence may require him/her to submit to an eyesight test. Refusing to do so is a further offence under s. 96(3).

The specific requirements as to eyesight are set out at regs 72–73 and sch. 8 to the Motor Vehicles (Driving Licences) Regulations 1999 (SI 1999 No. 2864). These regulations have been amended several times. They *generally* require a potential driver, existing licence holder and participant in compulsory basic training courses for motor cycles to be able to read:

- characters 79 mm high and 57 mm wide
- on a registration mark
- fixed to a motor vehicle
- at 20.5 metres
- in good light (with the aid of corrective lenses if worn at the time).

However, one amendment to the regulations takes into account the changes to the dimensions of vehicle registration plate characters and, if the new, narrower characters (50 mm wide) are used, the relevant distance for the above test is generally 20 metres.

For this reason it would seem that an eyesight test under s. 96(2) ought to be carried out under the same conditions, though there is no direct authority on the point. If the intention of parliament in passing s. 96(2) was to improve road safety, it would perhaps have been more useful to have a power to test a driver's ability to see clearly under the conditions which prevailed when he/she was found to be driving (e.g. at night).

One source of a police officer's 'reasonable suspicion' under s. 96(2) might be the information code on the driver's photocard licence which will state whether the holder has any eyesight correction (**see para. 11.3**).

Inability to read the characters as set out in the regulations will amount to a prescribed disability for the purposes of s. 92(2) (see above).

The 1999 Regulations also set out the conditions under which the Secretary of State may require (under s. 94) an applicant for a licence to undergo a medical examination if he/she has convictions for drink/driving offences (for which, **see chapter 5**).

If a driving licence has been refused or revoked on medical grounds (including eyesight), the Secretary of State may serve a notice (under s. 94(5)(c) of the Road Traffic Act 1988) requiring the applicant to take a specific driving test in order to assess his/her fitness—a 'disability assessment test'. In such cases the person may be granted a disability assessment licence which is an authority to drive only *for the purposes of taking such a test.*

The 1999 Regulations make special provision for drivers who suffer from epilepsy who may obtain a Group 1 licence as long as they have been free from an attack over the preceding 12 months or have only suffered from those attacks while asleep.

A person refused a licence may appeal against that decision under s. 100.

A court has a duty to notify the Secretary of State if it appears that a person has a relevant disease or disability (s. 22 of the Road Traffic Offenders Act 1988).

The Secretary of State may attach conditions to a licence in light of any disability of the holder. If the holder does not observe those conditions, he/she commits the offence under s. 87(1) (see para. 11.3.3).

Section 94 imposes a requirement for a licence holder to notify the Secretary of State *in writing*, of any relevant disability which has either not been disclosed in the past or which has become more acute since the licence was granted.

OFFENCE: **Failing to Give Notification of Relevant Disability—**
Road Traffic Act 1988, s. 94(3)

- Triable summarily • Fine

(No specific power of arrest)

The Road Traffic Act 1988, s. 94 states:

(3) A person who fails without reasonable excuse to notify the Secretary of State as required by subsection (1) above is guilty of an offence.

OFFENCE: **Driving Motor Vehicle before Giving Notification of Disability—**
Road Traffic Act 1988, s. 94(3A)

- Triable summarily • Fine • Discretionary disqualification

(No specific power of arrest)

The Road Traffic Act 1988, s. 94 states:

(3A) A person who holds a licence authorising him to drive a motor vehicle of any class and who drives a motor vehicle of that class on a road is guilty of an offence if at any earlier time while the licence was in force he was required by subsection (1) above to notify the Secretary of State but has failed without reasonable excuse to do so.

OFFENCE: **Driving Motor Vehicle after Refusal or Revocation—**
Road Traffic Act 1988, s. 94A

- Triable summarily • Six months' imprisonment and/or a fine • Discretionary disqualification

(No specific power of arrest)

The Road Traffic Act 1988, s. 94A states:

(1) A person who drives a motor vehicle of any class on a road otherwise than in accordance with a licence authorising him to drive a motor vehicle of that class is guilty of an offence if—
 (a) at any earlier time the Secretary of State—
  (i) has in accordance with section 92(3) of this Act refused to grant such a licence,
  (ii) has under section 93(1) or (2) of this Act revoked such a licence, or
  (iii) has served notice on that person in pursuance of section 99C(1) or (2) of this Act requiring him to deliver to the Secretary of State a Community licence authorising him to drive a motor vehicle of that or a corresponding class, and
 (b) since that earlier time he has not been granted—
  (i) a licence under this Part of this Act, or
  (ii) a Community licence, authorising him to drive a motor vehicle of that or a corresponding class.

11.7 The Road Traffic (Driver Licensing and Information Systems) Act 1989

The purpose of this Act is to provide a unified information system about and for drivers, under the authority of the Secretary of State.

Part I of the 1989 Act introduces a unified licensing system, abolishing the former special licences for heavy goods vehicles (HGV) and public service vehicles (PSV).

The Act provides for the licensing of drivers of such vehicles and sets out the requirements of those drivers, including their suitability to hold the relevant licence.

The Act provides the Secretary of State with powers to refuse driving licences to some applicants or to revoke existing licences on grounds of physical unfitness.

Part II of the Act provides for the introduction of driver information systems which will collect, store, process and transmit data on 'driver information'. Such systems are intended to help drivers in relation to traffic routes, congestion, etc. The systems are to be provided by operators licensed to do so by the Secretary of State. Operating an unlicensed system is an offence (s. 9).

11.7.1 Access to Driver Licensing Records

Under s. 71(2) of the Criminal Justice and Court Services Act 2000, the Secretary of State may make available any information held by him (e.g. by the DVLA) under Part III of the Road Traffic Act 1988 to the Police Information Technology Organisation (PITO). When making this information available to PITO, the Secretary of State may determine the *purposes* for which constables may be given that information and also the *circumstances* under which constables may further disclose the information they have been given.

In the Motor Vehicles (Access to Driver Licensing Records) Regulations 2001 (SI 2001/3343), the Secretary of State has set out both the purposes for which the information may be given to police officers by PITO and the circumstances under which officers can further disclose it.

Under reg. 2 of the 2001 Regulations, the purposes are the prevention, investigation or prosecution of a contravention of any provision under the following:

- the Road Traffic Act 1988
- the Road Traffic Offenders Act 1988

and also for ascertaining whether a person has had an order made in relation to them under the various statutes that allow for their disqualification from driving (namely the Child Support Act 1991, s. 40B and the Crime (Sentences) Act 1997, ss. 39(1) and 40(2)).

Under reg. 3, the circumstances under which officers may further disclose the information that they have been given are:

- where the information is passed to an employee of a police authority
- for any purpose ancillary to, or connected with, the use of the information by the officers.

11.8 New Drivers

The Road Traffic (New Drivers) Act 1995 places additional requirements on those who have recently 'qualified' as drivers. It sets out a 'probationary' period of two years and if, during that period, a driver receives six or more penalty points on his/her licence, the full entitlement to drive will be lost (s. 2). The driver concerned must then pass a further test of competence in the category of vehicle which he/she was entitled to drive.

Despite this legislation, there has been some continued concern at the standards of learner and new drivers. According to the Department for Transport, Local Government and the Regions, as many as one in five drivers have accidents within their first 12 months of obtaining a licence. At the time of writing, a consultation report had been published and further legislation including compulsory training, the use of special plates for newly-qualified drivers and more stringent requirements relating to speed, driving times and even blood-alcohol levels was being debated.

12 Excise and registration

12.1 Introduction

The law regulating the registration of vehicles and the charging of excise duty on them was largely consolidated in the Vehicle Excise and Registration Act 1994. As with most other areas of road traffic law however, the main Act is supported by a substantial number of statutory instruments making further regulations.

In addition, as one of the purposes behind the legislation is to generate income, the law is regularly extended and amended by the various Finance Acts which should be consulted in each case. The Greater London Authority Act 1999 creates special powers to levy charges for keeping or using motor vehicles on roads in London.

Much of the responsibility for managing the provisions of the 1999 Act fall to the Driver and Vehicle Licensing Agency (DVLA). The Criminal Justice and Court Services Act 2000 contains provisions to allowing increased police access to DVLA databases.

Both these well-established regulatory areas provide an opportunity for controlling and tracking the transfer and alteration of vehicles. By imposing further requirements on those who buy, sell, transfer and change vehicles, along with additional responsibilities for reporting and recording the supply of registration plates, the government has tightened up the controls on vehicle transfers and transactions through the Vehicles (Crime) Act 2001. Those controls are contained in a range of substantive and procedural amendments to the process of vehicle registration and licensing. In addition to the imposition of new statutory frameworks on people such as suppliers of vehicle number plates and motor salvage operators, the changes mean that every motorist wishing to buy a tax disc for their vehicle must now show their renewal notice (V11) or their registration document (V5). There are also more stringent requirements in relation to vehicles that are 'written off'.

12.2 Categories and Exemptions

12.2.1 Rates of Duty

The relevant rate of excise duty must be paid by anyone using or keeping a mechanically propelled vehicle on a public road unless the vehicle is exempt. The relevant rates of duty can be found in sch. 1 to the Vehicle Excise and Registration Act 1994 and also in the latest Finance Act.

Note that excise duty has to be paid on everything which, whether a vehicle or not, has been but has now ceased to be, a mechanically propelled vehicle and is either registered under the Act or is unregistered but is used or kept on a public road. This is a very wide definition.

The 1994 Act provides for vehicles to be 'taxed' at different rates according to their category and it sets out many different categories of vehicle, each of which is defined within s. 62 and sch.1.

12.2.2 Exemptions

Certain vehicles are exempt from excise duty. These are set out in sch. 2, as amended, and include:

- vehicles used for police, ambulance or fire brigade purposes
- mines rescue and lifeboat haulage vehicles
- certain vehicles used for the carriage of disabled people
- electrically assisted pedal cycles
- vehicles neither made nor adapted to carry the driver or passengers
- some agricultural or forestry vehicles
- certain vehicles when being tested
- vehicles temporarily in the country or to be exported.

Part VI of the Road Vehicles (Registration and Licensing) Regulations 2002 (SI 2002/2742) makes a number of exemptions for Crown vehicles and the Regulations also exempt vehicles that are imported by members of visiting forces and their dependants (see sch. 5).

Vehicles which are more than 25 years old may qualify as 'exempt' from excise duty provided they were constructed before 1 January 1973 (see the Finance Act 1998, s. 17). Even though they do not need to have duty paid on them, exempt vehicles may still be required to have a 'nil' licence and to display it in the appropriate manner when used on a public road (**see para. 12.6.3**).

12.2.3 The Main Offences

OFFENCE: **Using or Keeping an Unlicensed Vehicle on a Public Road—*Vehicle Excise and Registration Act 1994, s. 29***

• Triable summarily • Fine plus back duty

(No specific power of arrest)

The Vehicle Excise and Registration Act 1994, s. 29 states:

(1) If a person uses, or keeps, on a public road a vehicle (not being an exempt vehicle) which is unlicensed he is guilty of an offence.

(2) For the purposes of subsection (1) a vehicle is unlicensed if no vehicle licence or trade licence is in force for or in respect of the vehicle.

KEYNOTE

'Public roads' are roads maintained at the public expense, a much narrower definition than that of 'road' generally (**see para. 1.2.7**).

For 'using', **see chapter 1**.

'Vehicle' means a mechanically propelled vehicle or anything (whether or not it is a vehicle) that has been, but has ceased to be, a mechanically propelled vehicle (s. 1B).

'Exempt' vehicles may be issued with a 'nil' licence under the Finance Act 1997; if so, using or keeping them without such a licence being in force will be an offence under the specific offence created under s. 43A of the 1994 Act and not the above offence.

A keeper of a vehicle may make a statutory declaration under s. 22 to the effect that a vehicle will not be used or kept on a public road during a specified period. This declaration is known as a Statutory Off-Road Notification (SORN) and particulars for such a notification can be demanded by the Secretary of State from keepers of vehicles who do not renew their licences.

Proceedings under s. 29 may only be brought by the Secretary of State or by the police with his or her approval (s. 47(1)). A certificate in the approved form giving that authority is sufficient proof of that authority being given (s. 47(4)).

Full details of how to make a SORN are set out in reg. 26 and sch. 4 to the Road Vehicles (Registration and Licensing) Regulations 2002 (SI 2002/2742). Making a false SORN is a summary offence punishable by a fine (see sch. 8 to the Regulations).

In proving an offence under s. 29(1), evidence obtained via an approved device using automatic number plate recognition is admissible (under s. 20 of the Road Traffic Offenders Act 1988).

Under sch. 2A the Secretary of State may provide for the wheel-clamping, removal and disposal of vehicles where an offence under s. 29(1) above is being committed (see the Vehicle Excise Duty (Immobilisation, Removal and Disposal of Vehicles) Regulations 1997 (SI 1997/2439)).

The 1997 Regulations empower 'authorised persons' (who may be employed by local authorities, members of a police force or any other person authorised by the Secretary of State (reg. 3)) to clamp certain vehicles. Generally, clamping may take place when the authorised person has reason to believe that an offence under s. 29 of the Vehicle Excise and Registration Act 1994 involving a vehicle which is stationary on a public road is being committed (reg. 5).

The vehicle may be clamped where it is at the time or it may be moved to another place on a public road and clamped there.

The Regulations go on to provide a number of summary offences in relation to the unauthorised removal of, or interference with immobilisation notices and devices (reg. 7). Regulations 8, 13 and 15 create either way offences, punishable with a maximum of two years' imprisonment on indictment for making false declarations to secure the release of a clamped vehicle or to show that such a vehicle is 'exempt' (see below).

Under reg. 10 the vehicle may be disposed of by the relevant authority. The time within which this disposal may take place depends on the vehicle's condition. Vehicles of 'no economic value' are those where the total charges payable after seven days' storage will be more than the resale or scrap value of the vehicle. Vehicles falling into this category cannot be disposed of before seven days beginning with the time when they were removed; in the case of any other vehicles this period is 14 days.

Exemptions

Under reg. 4 the following vehicles are generally exempt from the provisions of the 1997 Regulations when being used on a public road:

- Vehicles displaying a **current** disabled person's badge — C
- Vehicles stationary at a time when **less** than 24 hours have elapsed since their release from immobilisation/removal — L
- Vehicles that appear to have been **abandoned** — A
- Vehicles displaying a British **Medical** Association car badge — M
- PSVs being used for the carriage of **p**assengers; vehicles being used by a **p**ublic utility or the **p**ost office — P
- Vehicles **exempt** from excise duty **displaying** nil licences — E D

The above list is a summary of the exemptions. For the specific conditions under which exemption is granted, see reg. 4.

A defendant may be required to pay up to five times the annual rate of duty applicable to the particular type of vehicle (s. 29(3)) and must pay any back duty.

The offence is committed at the time at which the vehicle is shown to have been used or kept on a public road. The practice of allowing 14 days' grace does not affect that situation, neither does the fact that a defendant buys a licence immediately afterwards (see *Wharton* v *Taylor* (1965) 109 SJ 475).

The fact that 'the cheque is in the post' will not save a defendant from the provisions of s. 29 either (*Nattrass* v *Gibson* (1968) 112 SJ 866).

The burden of proof in relation to the character, weight or cylinder capacity of the vehicle used, or of the purpose to which the vehicle was being put lies with the defendant for this offence (s. 53).

OFFENCE: **Failing to Display Licence—*Vehicle Excise and Registration Act 1994, s. 33(1) and (1A)***

• Triable summarily • Fine

(No specific power of arrest)

The Vehicle Excise and Registration Act 1994, s. 33 states:

(1) A person is guilty of an offence if—
  (a) he uses, or keeps, on a public road a vehicle in respect of which vehicle excise duty is chargeable, and
  (b) there is not fixed to and exhibited on the vehicle in the manner prescribed by regulations made by the Secretary of State a licence for, or in respect of, the vehicle which is for the time being in force.

(1A) A person is guilty of an offence if—
  (a) he uses, or keeps, on a public road an exempt vehicle,
  (b) that vehicle is one in respect of which regulations under this Act require a nil licence to be in force, and
  (c) there is not fixed to and exhibited in the vehicle in the manner prescribed by regulations made by the Secretary of State a nil licence for that vehicle which is for the time being in force.

KEYNOTE

Although it is not common practice to prosecute both offences in full, there is no reason why both failing to display a current licence and using/keeping a vehicle without having such a licence in force cannot be charged together (*Pilgram* v *Dean* [1974] RTR 299).

The 'manner prescribed' is set out in reg. 6 of the Road Vehicles (Registration and Licensing) Regulations 2002 and, in the case of any vehicle fitted with a glass windscreen in front of the driver extending across the vehicle to its near side, on or adjacent to the near side of the windscreen so that *all the particulars on the licence* are clearly visible in daylight from the near side of the road. Other provisions are made for vehicles not within this description.

Regulation 16(1) of the 1971 Regulations allows for a very limited defence where the person removes the licence at a post office to take the old one out and replace it.

The offence appears to be an 'absolute' liability offence (see Crime, chapter 1), and no guilty state of mind is needed; it may even be committed if someone else removes the licence from the vehicle (see *Strowger* v *John* [1974] RTR 124).

The Secretary of State may also make regulations prohibiting a person from exhibiting on a vehicle which is kept on a public road anything that is intended to be, or that could reasonably be mistaken for a current licence (s. 33(4)).

There are various other offences connected with being the owner of a registered vehicle that is unlicensed, along with numerous exceptions, conditions and defences (see generally ss. 31 and 31A).

OFFENCE: **Not Paying Duty Chargeable at Higher Rate—*Vehicle Excise and Registration Act 1994, s. 37(1)***

• Triable summarily • Fine

(No specific power of arrest)

The Vehicle Excise and Registration Act 1994, s. 37 states:

(1) Where—
  (a) a vehicle licence has been taken out for a vehicle at any rate of vehicle excise duty,
  (b) at any time while the licence is in force the vehicle is so used that duty at a higher rate becomes chargeable in respect of the licence for the vehicle under section 15, and
  (c) duty at that higher rate was not paid before the vehicle was so used,

the person so using the vehicle is guilty of an offence.

KEYNOTE

If a vehicle is taxed at a certain rate and is then used in a way that attracts a higher rate, the higher duty becomes payable (s. 15). Failing to pay that higher rate is an offence which can be punished by a penalty of five times the difference between the rate paid and the annual higher rate. Using a vehicle in breach of certain conditions attached to a low rate licence would be an example of where this might happen.

As with the offences of failing to display the correct excise licence (above), the burden of proof in relation to the character, weight or cylinder capacity of the vehicle used, or of the purpose to which the vehicle was being put lies with the defendant for this offence (s. 53).

12.3 Notification of Changes

The whole system imposing obligations on owners of vehicles to notify the Secretary of State of various changes and transactions is now set out under the Road Vehicles (Registration and Licensing) Regulations 2002 (SI 2002/2742). Part III deals with the issuing and production of registration documents.

Under reg. 12, the keeper of a vehicle must produce the registration document for inspection at any reasonable time by a constable or person acting on behalf of the Secretary of State. There is also a series of requirements that surround the application for a replacement registration document or the issue of a new one (see regs 13–15).

Part IV of the Regulations deals with the notification requirements where alterations are made to a vehicle, including a change of name or address by the keeper and the destruction or permanent export of the vehicle. In particular, reg. 15(3) and sch. 3 set out the requirements where vehicles are being written off by insurers. Regulations 20–25 set out the general provisions that govern the notification requirements when there is a change in the keeper of a vehicle and these requirements vary depending on whether the registration document was issued in Great Britain before 24 March 1997 or not. Breaches of specific provisions (listed at sch. 8) will generally be summary offences under s. 59(2) of the Vehicle Excise and Registration Act 1994, punishable by a fine.

Part V of the Regulations allows for the Secretary of State to disclose any particulars from the register to a number of people including chief officers of police; it also allows the Secretary of State to sell some of the information recorded there.

12.4 Trade Licences

Motor traders and vehicle testers may apply for trade licences which exempt them, under certain circumstances, from the rates of duty applicable above (s. 11 of the Vehicle Excise and Registration Act 1994).

The detail for the various obligations placed on trade licence holders are to be found in Part VII of the Road Vehicles (Registration and Licensing) Regulations 2002 (SI 2002/2742).

These begin by describing the business of a motor trader as defined under s. 62 of the Vehicles Excise and Registration Act 1994. In summary this covers anyone in the business of modifying vehicles, whether by the fitting of accessories or otherwise and the business of valeting vehicles. Thus a motor trader is involved in:

- manufacturing
- repairing

- dealing in
- modifying, or
- valeting vehicles.

'Dealing' includes those whose business is wholly or mainly concerned with collecting or delivering vehicles (s. 62(1)).

Conditions of Issue

The conditions under which trade licences are to be issued can be found in part I of sch. 6 to the Regulations. These are summarised below.

If the holder of a trade licence changes their name, the name of their business or their business address, they must notify the change and the new name or address to the Secretary of State and send the licence to the Secretary of State for any necessary amendment.

The holder of the licence must not (or permit anyone else to):

- alter, deface, mutilate or add anything to a trade plate;
- exhibit on any vehicle any trade licence or trade plate that has been altered, defaced or added to or on which the characters have become illegible or the colour changed;
- exhibit on any vehicle anything which could be mistaken for a trade plate;
- display the trade licence or any trade plates on any vehicle unless that vehicle is within the relevant class specified in s. 11(2) of the 1994 Act (if the holder is a motor trader who is a manufacturer of vehicles), s. 11(3) (if the holder is any other motor trader), or s. 11(4) (if the holder is a vehicle tester), and the vehicle is being used for one or more of the prescribed purposes in accordance with reg. 36 and sch. 6 (see below).

In addition, the holder of the licence must not permit any person to display the trade licence or any trade plates on a vehicle except a vehicle which that person is using for the purposes of the holder under the licence, nor must he/she display any trade plate on a vehicle used under the licence unless the trade plate shows the *general registration mark* assigned to the holder in respect of that licence.

Purposes of Issue

The purposes for which a trade licence can be used on a public road vary depending on whether the licence holder is a motor trader or a vehicle tester. They are set out fully in part II of sch. 6 but are summarised as follows:

Generally a vehicle being used on a public road under a trade plate cannot carry goods or burdens of any description. However, they may carry 'specified loads'. Listed at para. 13 of sch. 6, these loads include loads which are built in as part of the vehicle or permanently attached to it or, in some cases, loads consisting of a trailer or of parts, accessories or equipment designed to be fitted to the vehicle and of tools for fitting them. Test loads carried solely for the purpose of testing or demonstrating the vehicle or any of its accessories or equipment are also permitted under certain circumstances.

Similarly, vehicles used under trade plates cannot carry passengers except a person carried in connection with the relevant business purpose (e.g. a prospective buyer).

Motor traders may use a vehicle (other than a vehicle kept for research and development) for business purposes as a manufacturer or repairer of, or dealer in vehicles and trailers; for modifying vehicles (whether by the fitting of accessories or otherwise) or for valeting vehicles.

Motor traders may also use a vehicle under a trade licence for what are termed 'paragraph 12 purposes'. Paragraph 12 purposes are all the things that you would expect a motor trader to have to do necessarily in the course of his/her business, such as:

- road testing vehicles, accessories or equipment in the ordinary course of construction, modification or repair or after completion;
- road testing vehicles for the benefit of a prospective purchaser and going to or from any place for the purpose of such test or trial;
- proceeding to or from a public weighbridge or to or from any place for registration or inspection by a person acting on behalf of the Secretary of State, or to or from a place where the vehicle is to be or has been inspected or tested;
- for delivering a vehicle to the place where the purchaser intends to keep it or for demonstrating its operation (or the operation of its accessories or equipment) when it is being handed over to the purchaser;
- for delivering a vehicle from one part of the licence holder's premises to another or between parts of premises of another manufacturer, repairer or dealer in vehicles, or for proceeding to a place where it is to be broken up or otherwise dismantled.

Vehicle testers are people, other than motor traders, who regularly in the course of their business test vehicles belonging to other people on roads (s. 62(1)).

A vehicle tester may use a vehicle on a public road under a trade licence for the purposes of testing it (or any trailer drawn by it) or any of the accessories or equipment on the vehicle or trailer in the course of the business of the holder of the trade licence as a vehicle tester. The restrictions on carrying goods or burdens under a vehicle tester's licence are greater than those applying to motor traders (see sch. 6, para. 15(2)).

Trade Plates

Every holder of a trade licence will be issued with a set of trade plates appropriate to the class of vehicles for which the licence is to be used (reg. 40(1)). Each trade plate will show the general registration mark assigned to the holder, and one of the trade plates must include a means whereby the licence may be fixed to it (reg. 40(2)). If the licence is only to be used in respect of motorcycles, the holder will only be issued with one trade plate (which, of course, will have somewhere to fix the licence on it).

When a vehicle is being used under a trade licence, trade plates must be fixed and displayed and illuminated in a way which complies with the Road Vehicles (Display of Registration Marks) Regulations 2001 (SI 2001/561)—that is, broadly they must be clearly displayed on the front and the back of the vehicle.

Breaches of the specific provisions in sch. 8 (failing to return trade plates and irregularities in displaying them) are summary offences under s. 59(2) of the Vehicle Excise and Registration Act 1994, punishable by a fine.

OFFENCE: **Improper Use of Trade Licence—*Vehicle Excise and Registration Act 1994, s. 34(1)***

- Triable summarily • Fine

(No specific power of arrest)

The Vehicle Excise and Registration Act 1994, s. 34 states:

(1) A person holding a trade licence or trade licences is guilty of an offence if he—

(a) uses at any one time on a public road a greater number of vehicles (not being vehicles for which vehicle licences are for the time being in force) than he is authorised to use by virtue of the trade licence or licences,

(b) uses a vehicle (not being a vehicle for which a vehicle licence is for the time being in force) on a public road for any purpose other than a purpose which has been prescribed under section 12(2)(b), or

(c) uses the trade licence, or any of the trade licences, for the purposes of keeping on a public road in any circumstances other than circumstances which have been prescribed under section 12(1)(c) a vehicle which is not being used on that road.

KEYNOTE

Section 34 creates three different offences.

A licence holder may be required to pay up to five times the duty payable on the particular class of vehicle concerned.

Only one vehicle may be used on one licence at a time but a person may hold more than one trade licence.

The burden of proof in relation to the character, weight or cylinder capacity of the vehicle used, the number of vehicles used or of the purpose to which the vehicle was being put lies with the defendant for this offence (s. 53).

12.4.1 Dishonoured Cheques

Vehicle licences and trade licences may be paid for by cheque (s. 19A(1) of the 1994 Act). If a cheque used to pay for either type of licence is subsequently dishonoured, the Secretary of State (through the DVLA) may send a notice to the licence holder informing him/her that the licence is void as from the time it was granted (s. 19A(2)). Where such a notice is sent, the licence holder may also be required to deliver up that licence. If the licence holder fails to do so, he/she commits a summary offence punishable by a fine or an amount equal to five times the duty payable on that vehicle (s. 35A). The court must also impose a requirement to pay any outstanding duty owed on the vehicle since the application was made or since it was due to take effect (s. 36).

12.5 Furnishing Information

Section 46 of the Vehicle Excise and Registration Act 1994 states:

(1) Where it is alleged that a vehicle has been used on a road in contravention of section 29, 34 or 37—
   (a) the person keeping the vehicle shall give such information as he may be required to give in accordance with subsection (7) as to the identity of the driver of the vehicle or any person who used the vehicle, and
   (b) any other person shall give such information as it is in his power to give and which may lead to the identification of the driver of the vehicle or any person who used the vehicle if he is required to do so in accordance with subsection (7).

(2) Where it is alleged that a vehicle has been kept on a road in contravention of section 29—
   (a) the person keeping the vehicle shall give such information as he may be required to give in accordance with subsection (7) as to the identity of the person who kept the vehicle on the road, and
   (b) any other person shall give such information as it is in his power to give and which may lead to the identification of the person who kept the vehicle on the road if he is required to do so in accordance with subsection (7).

(3) Where it is alleged that a vehicle has at any time been used on a road in contravention of section 29, the person who is alleged to have so used the vehicle shall give such information as it is in his power to give as to the identity of the person who was keeping the vehicle at that time if he is required to do so in accordance with subsection (7).

OFFENCE: **Failing to Give Information—*Vehicle Excise and Registration Act 1994, s. 46(4)***

- Triable summarily • Fine

(No specific power of arrest)

The Vehicle Excise and Registration Act 1994, s. 46 states:

(4) A person who fails to comply with subsection (1), (2) or (3) is guilty of an offence.

KEYNOTE

The requirement for information to be given must come on behalf of a chief officer of police or the Secretary of State (s. 46(7)).

The obligation to furnish the required information, in relation to s. 46(1)(a) and (2)(a), applies to the person who is the keeper *at the time of the requirement*, even if they were not the keeper at the time of the alleged offence (*Hateley* v *Greenough* [1962] Crim LR 329).

Section 46A goes on to make further provisions for the Secretary of State to require information leading to the identity of a person who is keeping or who has sold or disposed of a vehicle in contravention of the regulations relating to the registration of vehicles (i.e. under s. 22). Failing to comply with a notice under s. 46A is an offence similar to that under s. 46(4) above.

This offence also applies to the offence relating to 'nil licences' under s. 43A (**see para. 12.2.3**).

Evidence that a person has been served with a notice requiring the relevant information is admissible in proving the listed offences (s. 51), so too is evidence held by the Secretary of State (DVLA) in official records (s. 52).

Section 45 creates a number of offences concerned with the making of false or misleading statements made in connection with just about any transaction, application or duty to furnish information under the Act.

12.5.1 Defence

Section 46(6) states:

> (6) If a person is charged with an offence under subsection (4) consisting of failing to comply with subsection (1)(a) or (2)(a), it is a defence for him to show to the satisfaction of the court that he did not know, and could not with reasonable diligence have ascertained, the identity of the person or persons concerned.

12.6 Registration Marks

Sections 21 to 28 of the Vehicle Excise and Registration Act 1994, as amended, contain the provisions relating to the registration of vehicles.

Section 21 makes detailed provision for the requirements to provide information when registering, selling and disposing of vehicles. It also places certain obligations on people who surrender vehicle licences or who do not renew them (see s. 22(1C), (1D) and (1E)). The applicable regulations governing these aspects of registration are the Road Vehicles (Registration and Licensing) Regulations 2002 (SI 2002/2742).

Under s. 23, the Secretary of State must assign a registration mark to a vehicle when it is registered.

The main regulations relating to registration marks are the Road Vehicles (Display of Registration Marks) Regulations 2001 (SI 2001/561). Parts II and IV impose conditions as to the manner and form in which registration plates must be exhibited. Broadly, invalid vehicles and some pedestrian controlled-vehicles are exempt from the need for registration plates.

The rules governing the form and layout of registration marks have been extended and now appear in the Road Vehicles (Display of Registration Marks) Regulations 2001. The increasingly common use of stylised lettering on number plates was not formerly unlawful, allowing for the creation of many elaborate but illegible plates. Under the 2001 Regulations, number plates must conform to the specifications set out in schs 2 and 3. Regulation 15 of the 2001 Regulations provides that registration plates fixed to vehicles first registered on or after 1 September 2001 must have characters in the prescribed font only and that plates on other vehicles must be in a font or style which is 'substantially similar' to that font and are easily distinguishable. Generally, characters must not be formed:

- in italic script
- using a font in which the characters are not vertical
- using multiple or broken strokes or
- in such a way as to make a character(s) appear like a different character(s).

In addition, reg. 11 prohibits the use of screws, bolts or fixing devices which have the effect of changing the appearance or legibility of any of the characters of the registration mark.

There are some 'special cases' provided for under reg. 14A but these are just that—they are not generally production line cars that appear every day on the roads in England and Wales.

These provisions came into force on 1 September 2001. So for many tortuously styled number plates, these new Regulations may well spell T 1•1 3 END.

Regulation 11 also prohibits the use of reflex-reflecting material, retro-reflective characters and any other treatment of the number plate which renders the characters less easily distinguishable to the eye *or* which would impair the making of a true photographic image of the plate. This regulation is designed to prevent the proliferation of plates that cannot be picked up by traffic regulation cameras.

The permitted layout, spacing and required character sizes are set out in reg. 14 and sch. 3 to the 2001 Regulations. These are very detailed and make fine distinctions between different types of vehicles, for instance, the required size of characters on the number plates of motor cycles, motor tricycles and quadricycles is different from that of cars.

There are now two widths of number plate characters, with the newer ones being slightly narrower than the others. One practical effect of this for the police is that the permitted distances from which drivers must be able to read such characters are reduced by 0.5 m (**see chapter 11**).

Regulation 16 prohibits the display of any material other than a registration mark on a number plate other than 'international distinguishing signs' (e.g. the GB sign in the left-hand side of some dual purpose number plates). These regulations have also been amended to allow for national flags to be displayed in the same place on the number plate as the other distinguishing signs. This will allow for the Union flag, the cross of St George, the Scottish Saltire and the Welsh dragon to be displayed, along with the relevant initials (Eng, Cym, etc.).

The fixing of registration plates to vehicles is now governed by part II of the 2001 Regulations. Part II generally applies to relevant vehicles other than works trucks, agricultural machines and road rollers. These vehicles must have a registration plate fixed to the rear of:

- the vehicle, or
- where it is towing a trailer, the trailer or rearmost trailer

(reg. 5).

Front registration plates are generally required by the same vehicles, the primary exception being motor cycles or motor tricycles (**see chapter 1**) which do not have bodies like a four-wheeled vehicle (e.g. the old Robin Reliant type car). A front registration plate *need not* be fixed to such motor cycles or motor tricycles if they were first registered before 1 September 2001; they *must not* be fixed to such vehicles registered on or after that date (reg. 6).

From 1 September 2001 works trucks, agricultural machines and road rollers must have registration plates and these can generally be fixed either on both sides of the vehicle or on the rear (reg. 8).

Generally, all number plates must be fixed vertically or as near to vertical as is reasonably practicable and in such a position that, in normal daylight, the characters are easily distinguishable from every part of the relevant area (see regs 5 to 8).

12.6.1 Supply of Registration Plates

Part 2 of the Vehicles (Crime) Act 2001 imposes statutory regulation on suppliers of vehicle registration plates by creating a number of obligations on suppliers and also by creating new offences connected with the supply of number plates.

Registration

Any person who carries on a business, which consists wholly or partly in selling registration plates, who is not an exempt person, has to be registered in accordance with the 2001 Act and the relevant regulations made under it. Carrying on such a business without being registered is a summary offence punishable by a fine (s. 17(1)).

Vehicle dealers who sell vehicles for which they have arranged first registration in the United Kingdom on behalf of the intending purchaser will be exempt from these provisions (see the Vehicles Crime (Registration of Registration Plate Suppliers) (England and Wales) Regulations 2002 (SI 2002/2977), reg. 3).

Under s. 18, the Secretary of State must establish and maintain a register of persons carrying on business as registration plate suppliers. Each person's entry in the register must contain the details prescribed in the 2002 Regulations above. Under s. 21, the Secretary of State may (subject to an appeal process) cancel a person's registration if satisfied that the person concerned is not carrying on business as a registration plate supplier and has not, while registered, been doing so for at least 28 days.

As with the register of motor salvage operators (as to which, **see para. 12.7**), the register must contain details from which the registered person or company can be identified, however the amount of detail required to be recorded here is less than that for salvage operators.

Under reg. 4 the register must include:

- in the case of a sole trader, their full name and usual residential address, along with any other name under which they trade and the address of any premises where they carry on that business;
- in the case of a company, its name and registered number, along with the address of its principal place of business and the address(es) of any other premises where it carries on that business;
- in the case of a partnership, its name and the address(es) of its principal place of business and any other premises where it carries on that business.

In all cases the register must contain a telephone number for the registered person, the registered identity number, the date of registration (and any suspension) and details of any conviction for an offence under Part 2 of the 2001 Act within the last five years. If the person concerned was under 18 at the time of conviction, then the relevant convictions are those occurring within the past 30 months.

The Secretary of State will (subject to exemptions) supply any person who has requested information contained in the register with the information that he/she has requested (s. 18(5)) on payment of the appropriate fee.

The Secretary of State may also make all of the information contained in the register available to the Police Information Technology Organisation (as to which, **see General Police Duties, chapter 1**) for use by constables for the purpose of investigating offences under Part 2 of the Act (s. 18(7)).

Before selling a registration plate, a person registered must obtain from the prospective purchaser the following information:

- the name and address of the purchaser (or the appropriate details from a company or partnership). If an individual is buying the plates for someone else, that other person's name and address must also be obtained.

  This information must be verified by the registered person against the *documents* set out in part I of the schedule to the Regulations (**see appendix 13**). Part I of the schedule is very specific about the combinations of documents that must be checked and includes utility bills, drivers' licences, passports, etc;
- the registration mark to be shown on the plates;
- if the plates are being fitted by a vehicle body repairer at the request of an insurance company, the name of that company and the relevant policy number;
- the connection of the purchaser with the registration mark or the vehicle on which it is to be fixed. This information must be verified by the registered person against *a document* set out in part II of the schedule to the Regulations (**see appendix 13**). Unlike part I of the schedule, part II only requires one of the specified documents to be checked (registration documents, etc.) and the requirement will not apply if the plates are being fitted by a vehicle body repairer at the request of an insurance company

(see generally reg. 6).

Contravening or failing to comply with these regulations will be a summary offence punishable with a fine (s. 25(3) and reg. 6(7)).

It is a defence for the person to show that he/she took all reasonable steps and exercised all due diligence to avoid committing this offence (s. 25(4)).

Records

Every person registered under s. 18 of the Act must keep records either at the principal place of business or any other premises where they carry on the business of registration plate supply (reg. 7).

The records must contain the following detail in relation to registration plates sold by that person and for three years thereafter:

- all the information required to be obtained under reg. 6 above;
- a description of the method by which the plates were paid for. If payment was by debit or credit card, the issuer of the card and the account number must be recorded. If a cheque was used, the name of the bank and the account number must be recorded;
- a statement of the relevant documents from the schedule used to verify the purchaser's name and address and connection with the vehicle (under reg. 6 above). If a driving licence was used, the driver number (or equivalent if non-UK) and reference/VIN numbers from any V5 document.

Contravening or failing to comply with these regulations will be a summary offence punishable with a fine (s. 24(4) and reg. 7(3)).

It is a defence for the person to show that he/she took all reasonable steps and exercised all due diligence to avoid committing this offence (s. 24(5)).

Making a statement which the person knows to be false in a material particular or making such statement recklessly when applying to be registered is a summary offence punishable by a fine (s. 19(3)). Where a person is convicted of this offence the court may, instead of or in addition to imposing a fine, order their removal from the register, prohibit them from making an application for registration for a period not exceeding five years or both (subject to an appeal process) (s. 20). Applying for registration while a court order is in force is a summary offence (s. 19(6)).

Where *a registered person* is convicted of any other offence under Part 2 of the Act, the court may, instead of or in addition to imposing a fine, suspend their registration for any period of up to five years (subject to an appeal process). As above, applying for registration while a court order suspending registration is in force is also a summary offence (s. 19(6)).

Powers of Entry and Inspection

As with the provisions relating to motor salvage operators (**see para. 12.7**), there is a power—although the Act refers to a 'right'—to enter premises under Part 2. However, unlike the motor salvage operators provisions, the power to enter applies to constables *and authorised persons*. Section 26 provides that a constable or person authorised by the local authority may, at any reasonable time, enter and inspect premises for the time being entered in the register as premises which are occupied by a person carrying on business as a registration plate supplier wholly or partly for the purposes of his/her business so far as it consists in selling registration plates. A constable or authorised person may also at any reasonable time:

- require production of, and inspect, any registration plates kept at premises above; and
- require production of, inspect and take copies of or extracts from any records which the person carrying on business as a registration plate supplier is required to keep at such premises by virtue of Part 2

(s. 26(2)).

Where, on application by a constable or an authorised person, a justice of the peace is satisfied that admission to premises specified in the application is reasonably required in order to secure compliance with the provisions of Part 2, or to ascertain whether those provisions are being complied with, the justice may issue a warrant authorising a constable or authorised person to enter and inspect the premises concerned (s. 26(3) and (4)).

The constable or authorised person, in seeking to enter any premises in the exercise of the above powers must, if required by or on behalf of the owner or occupier or person in charge of the premises, produce evidence of their identity, and of their authority for entering, before doing so (s. 26(6)). An authorised person seeking to enter under the authority of a warrant must also produce evidence of their identity.

In another similarity with Part 1, not only does Part 2 of the Act not give police officers a power to use force to enter premises—it specifically says that officers *will not be entitled* to use force (s. 26(5)). However, the 2001 Act goes on to say that reasonable force may be used if necessary when exercising the powers of entry under a warrant under s. 26 above.

It is a summary offence punishable by a fine to obstruct an authorised person (but not a constable as presumably the Police Act 1996 offence will cover this situation—**see Crime, chapter 8**) exercising their powers of entry with or without a warrant (s. 26(7)).

Notification

In order to keep the registers up to date, the 2001 Act imposes certain requirements on people who are applying to be registered or who are registered to notify the Secretary of State of any changes in their circumstances (see s. 27). A person applying to be registered must give notice of any changes affecting in a material particular the accuracy of the information, which they have provided in connection with their application. Once registered, a person must, within 28 days of the changes occurring, give notice of any changes affecting their entry in the register. A registered person who is not carrying on business as a registration plate supplier must give notice of that fact within 28 days of the beginning of the period in which they are not carrying on business while registered. Failure to give notice as required will be a summary offence punishable by a fine (s. 27(5)) but it is a defence to show that the person took all reasonable steps and exercised all due diligence to avoid committing the offence.

12.6.2 Lighting

The requirement to light the rear registration plate on a vehicle is covered under the Road Vehicles Lighting Regulations 1989 (**see chapter 9**).

The Road Vehicles (Display of Registration Marks) Regulations 2001 sets out the specific requirements of where and how such plates must be lit. Relevant vehicles (e.g. those other than works trucks, agricultural machines and road rollers) being used on a road between sunset and sunrise must have their rear registration plate lit so that it is easily distinguishable from the prescribed distance (15–18 metres) (reg. 9).

Strangely, the wording of reg. 9 is not as precise as that relating to the general legibility of registration plates (see above). Whereas the regulations governing general legibility require the *characters* themselves to be easily distinguishable, reg. 9 only requires the *plate itself* to be easily distinguishable—a minor point but one that will no doubt be made much of sooner or later.

12.6.3 Other Offences

OFFENCE: **Not Fixing Registration Mark—*Vehicle Excise and Registration Act 1994, s. 42(1)***

- Triable summarily • Fine

(No specific power of arrest)

The Vehicle Excise and Registration Act 1994, s. 42 states:

(1) If a registration mark is not fixed on a vehicle as required by virtue of section 23, the relevant person is guilty of an offence.

(2) ...

(3) In subsection (1) 'the relevant person' means the person driving the vehicle or, where it is not being driven, the person keeping it.

OFFENCE: **Obscured Registration Mark—*Vehicle Excise and Registration Act 1994, s. 43(1)***

- Triable summarily • Fine

(No specific power of arrest)

The Vehicle Excise and Registration Act 1994, s. 43 states:

(1) If a registration mark fixed on a vehicle as required by virtue of section 23 is in any way—
 (a) obscured, or
 (b) rendered, or allowed to become, not easily distinguishable,
 the relevant person is guilty of an offence.

KEYNOTE

The offence under each section is committed by the 'relevant' person; that is, the person *driving* the vehicle (**see chapter 1**) or, if the vehicle is not being driven, the keeper (s. 42(3)). A person is the 'keeper' if he/she causes the vehicle to be on a public road for any period, *however short*, when it is not in use there (s. 62(2)).

If the person can show that he/she had no reasonable opportunity to register the vehicle and that the vehicle was being driven for the purpose of it being so registered, he/she has a defence to a charge under s. 42(1); not fixing a registration mark (s. 42(4)). The burden of proof is upon the defendant.

In answer to a charge under s. 43(1) (obscured registration marks), the person has a defence if he/she can show that he/she took all steps which were reasonably practicable to prevent the mark being obscured or not easily distinguishable (s. 43(4)).

OFFENCE: **Failing to have Nil Licence for Exempt Vehicle—*Vehicle Excise and Registration Act 1994, s. 43A***

- Triable summarily • Fine

(No specific power of arrest)

The Vehicle Excise and Registration Act 1994, s. 43A states:

(1) A person is guilty of an offence if—
 (a) he uses, or keeps, on a public road an exempt vehicle,
 (b) that vehicle is one in respect of which regulations under this Act require a nil licence to be in force, and
 (c) a nil licence is not for the time being in force in respect of the vehicle.

KEYNOTE

A 'nil licence' means a document which is in the form of a vehicle licence and is issued by the Secretary of State in respect of an 'exempt vehicle' (s. 62).

The definition of 'nil licensable vehicles' can be found in the Road Vehicles (Registration and Licensing) Regulations 2002 (SI 2002/2742). Nil licences are renewable each year and are the same as a regular vehicle excise licence except that they have the word NIL marked in the space showing the amount of duty payable. Nil licences have to be displayed in the same manner as ordinary excise licences (see para. 12.3.3).

12.6.4 Overseas Vehicles

The Motor Vehicles (International Circulation) Regulations 1985 (SI 1985/610), as amended, govern the use of vehicles from abroad while in Great Britain.

Anyone who is resident outside Great Britain bringing a vehicle into the country must produce any registration card or document in respect of the vehicle to a police officer (or authorised person) on request at any reasonable time (reg. 4).

12.7 Motor Salvage Operators

Part 1 of the Vehicles (Crime) Act 2001 introduced a new statutory framework to regulate the motor salvage industry. In particular, every local authority has to establish and maintain a register in a prescribed form (s. 2) and every person who carries on a business as a motor salvage operator must register with the local authority. The definition of a motor salvage operator is set out in s. 1(2) and includes any person who carries on a business consisting:

(a) wholly or partly in the recovery for re-use or sale of salvageable parts from motor vehicles and the subsequent sale or other disposal for scrap of the remainder of the vehicles concerned;

(b) wholly or mainly in the purchase of written-off vehicles and their subsequent repair and re-sale;

(c) wholly or mainly in the sale or purchase of motor vehicles which are to be the subject (whether immediately or on a subsequent re-sale) of any of the activities mentioned in paragraphs (a) and (b);

. . .

'Motor vehicle' here means any vehicle whose function *is or was* to be used on roads as a mechanically propelled vehicle. 'Written-off motor vehicles' are motor vehicles which are in need of substantial repair but in relation to which a decision has been made not to carry out the repairs (see generally s. 16).

From 7 April 2003, vehicles that are written off as a result of an accident but which are later repaired, must undergo a Vehicle Identity Check (VIC) before they can be used on a road again. If any such vehicles are the subject of an insurance claim, the Motor Insurers' Anti-Fraud and Theft Register database will advise the DVLA. VICs will be carried out by Department of Transport vehicle inspectors at goods testing stations and the registration document of a vehicle that has undergone a VIC will have this fact recorded on it, together with the relevant date. Note, however, that this does not amount to evidence that a vehicle is in a roadworthy condition and should not be confused with the issuing of a test certificate (as to which, **see chapter 9**).

It is a summary offence punishable by a fine to carry on a business as a motor salvage operator in the area of a local authority without being registered for that area by the authority (s. 1(1)).

12.7.1 Registration

The particulars that must be set out in the local authority register are found in the Motor Salvage Operators Regulations 2002 (SI 2002/1916). They include:

- the full name and date of birth of the operator, all the directors (if a company) or the partners if relevant;
- of the operator is an individual or a partnership, the entry must show the home address of the individual or each partner. If the operator is a company, then the registered or principal office must be recorded; and
- each place in the local authority area which is occupied by the operator wholly, mainly or partly for the purpose of carrying on the business as an operator

(see reg. 3).

An operator will cease to be registered after three years unless they renew their registration (s. 2(4)). The details required for re-registration are set out in reg. 4 which requires, among other things, the disclosure of any convictions under Part 1 of the 2001 Act or of a 'specified offence' (see below).

The local authority must make sure that the contents of the register are available for inspection by members of the public at all reasonable times subject to such reasonable fees (if any) as the local authority may determine (s. 2(9)).

On receiving an application (and the appropriate fee), a local authority must register (or renew the registration of) the applicant unless it is satisfied that the applicant is not a fit and proper person to carry on business as a motor salvage operator (s. 3(3)). In deciding this issue, the local authority must have particular regard to whether the applicant has been convicted of any offences under Part 1 or a 'specified offence'.

A local authority can cancel a person's registration in the register for their area at any time if they are satisfied that the operator is not a fit and proper person to carry on business as such, or if satisfied that the operator is not carrying on business as such in their area *and has not, while registered, been doing so for at least 28 days*. There are, however, safeguards requiring the local authority to provide the operator with certain information and opportunities before cancelling their registration and there is also an appeal procedure before a magistrates' court (see generally ss. 4–6).

12.7.2 Records

Under s. 7 of the 2001 Act, a registered operator must keep certain records. The full detail of these requirements is set out in the 2002 Regulations. Such records may be kept in manual or electronic form. If they are manual records, they must be kept at the registered place of business, while electronic records must be accessible from that registered place of business (reg. 5).

The details required to be kept include all the things that common sense would require, including:

- details of the vehicle registration number, vehicle identification number (VIN), make, model and colour of the vehicle;
- the name, address and contact details of the supplier of the vehicle;
- the date on which the information above was entered on the record;
- [details of any proof of identity shown to the registered person by, or on behalf of the supplier of the vehicle, to establish the identity of the vehicle supplier;
- the general condition of the vehicle including details of the type of any damage to the vehicle (e.g. whether the damage has been caused by fire, water or impact) and the part of the vehicle damaged].

When a registered person sells (or otherwise disposes) of any vehicle, they must add the following pieces of information to the record above:

- the date of sale or other disposal of the vehicle;
- the name, address, and contact details of the person receiving the vehicle;
- the date when the information above was entered on the record;
- [details of any proof of identity shown to the registered person by, or on behalf of the purchaser of the vehicle, to establish the identity of the person receiving the vehicle;
- the condition of the vehicle at the time of the sale or other disposal (e.g. whether it was repaired, unrepaired, dismantled, or in the same condition as at purchase)].

The records required to be made under reg. 5 must be kept for a period of six years from the date of the last entry on the record for the vehicle.

A person who contravenes or fails to comply with any provision to which s. 7(4) applies commits a summary offence punishable by a fine. However, it should be noted that it is only a failure to record the details that are shown at the first three bullet points in each list above that will amount to an offence under s. 7(4)). This is why the others are shown in square brackets. Presumably, the incentive for recording the other details is that failure to do so will be taken into account by the local authority when considering a person's fitness to be registered as a motor salvage operator under s. 3.

A person who is registered in the register of a local authority shall give notice to the local authority of any changes affecting his/her entry in the register within 28 days of the changes occurring (s. 11).

The Secretary of State may make further regulations for the notification by registered persons of the destruction of motor vehicles (s. 8).

12.7.3 Powers of Entry and Inspection

Under s. 9 a constable may at any reasonable time enter and inspect premises for the time being entered in the register of a local authority as premises which are:

- occupied as a motor salvage yard by a person carrying on business as a motor salvage operator; or
- occupied by a person carrying on business as a motor salvage operator wholly or partly for the purposes of his/her business so far as it consists of any of the activities mentioned in s. 1(2) (see above).

A constable may also at any reasonable time:

- require production of, and inspect, any motor vehicles or salvageable parts kept at such premises; and
- require production of, inspect and take copies of or extracts from any records which the person carrying on business as a motor salvage operator is required to keep at such premises by virtue of Part 1 of the Act.

A 'motor salvage yard' means broadly any premises where any 'motor vehicles' (as defined above) are received or kept in the course of the carrying on of business as a motor salvage operator. However it does not include any premises where only salvageable parts of motor vehicles are so received or kept (see s. 16).

Any constable, in seeking to enter any premises in the exercise of the above powers must, if required by or on behalf of the owner or occupier or person in charge of the premises, produce evidence of his/her identity, and of his/her authority for entering, before doing so (s. 9(6)).

Note that, not only does Part 1 of the 2001 Act not give police officers a power to use force to enter premises—it specifically says that officers *will not be entitled* to use force (s. 9(5)). However, the Act goes on to say that reasonable force may be used if necessary when exercising the powers of entry under a s. 9 warrant. Therefore, if it is anticipated that the use of force will or may be necessary in securing entry, an application should be made for a warrant.

If, on application by a constable, a justice of the peace is satisfied that admission to premises specified is reasonably required in order to secure compliance with the provisions of Part 1 of the Act, or to ascertain whether those provisions are being complied with, he/she may issue a warrant authorising a constable to enter and inspect the premises concerned (s. 9(3) and (4)).

Making a statement which the person knows to be false in a material particular or making such statement recklessly when applying to be registered or to renew a registration is a summary offence punishable by a fine (s. 10(1)).

It is a further summary offence punishable by a fine for any person, on selling a motor vehicle to a motor salvage operator, to give that person a false name or address (see s. 12).

13 Goods and passenger vehicles

13.1 Introduction

The law regulating the use of goods and passenger vehicles is notoriously complex, not least because of the need to achieve a common transport policy in line with the UK's membership of the European Union.

The use and operation of goods and passenger vehicles, particularly in the course of a business, presents many opportunities for the creation of risk and danger to the community and is therefore seen as an area which requires fairly stringent legal provision. This chapter is intended as a summary of some of the more relevant of those provisions. The legislation which follows relies heavily on certain documents being held or produced by the relevant people; for the offences involving fraudulent use of such documents, **see chapter 16**.

13.2 Large Goods Vehicles and Passenger Carrying Vehicles

13.2.1 Large Goods Vehicles

A large goods vehicle (LGV) is:

- a motor vehicle (not being a medium-sized goods vehicle)
- which is constructed or adapted
- to carry or haul goods
- having a permissible maximum weight over 7.5 tonnes

(Road Traffic Act 1988, s. 121).

13.2.2 Passenger Carrying Vehicles

A passenger-carrying vehicle (PCV) is either:

- a vehicle used for carrying passengers
- which is constructed or adapted
- to carry more than 16 passengers (a 'large PCV')

or

- a vehicle used for carrying passengers *for hire or reward*
- which is constructed or adapted
- to carry more than 8 but not more than 16 passengers (a 'small PCV')

(Road Traffic Act 1988, s. 121).

13.2.3 LGV and PCV Licences

The driving licences under Part III of the Road Traffic Act 1988 (**see chapter 11**) are needed to drive large goods vehicles and passenger carrying vehicles. However, in order to drive such vehicles, a person must also hold an LGV or PCV licence granted under s. 110 of the 1988 Act.

The Motor Vehicles (Driving Licences) Regulations 1999 have been amended to permit the grant of an LGV licence to a person who is under 21 and who has accumulated not more than three penalty points, provided that he/she produces evidence of satisfactory completion of the off-road element of the Young LGV Driver Training Scheme.

If a person is granted an LGV or PCV licence, it is an offence to fail to comply with the conditions attached to it.

OFFENCE: **Conditions of LGV or PCV Licence—*Road Traffic Act 1988, s. 114***

- Triable summarily • Fine

(No specific power of arrest)

The Road Traffic Act 1988, s. 114 states:

(1) The following licences, that is to say—
 (a) a large goods vehicle or passenger-carrying vehicle driver's licence issued as a provisional licence,
 (b) a full large goods vehicle or passenger-carrying vehicle driver's licence granted to a person under the age of 21, and
 (c) a LGV Community licence held by a person under the age of 21 who is normally resident in Great Britain,
 shall be subject to the prescribed conditions, and if the holder of the licence fails, without reasonable excuse, to comply with any of the conditions he is guilty of an offence.

(2) It is an offence for a person knowingly to cause or permit another person who is under the age of 21 to drive a large goods vehicle of any class or a passenger-carrying vehicle of any class in contravention of the prescribed conditions to which that other person's licence is subject.

KEYNOTE

If a person drives an LGV or a PCV on a road without such a licence, they commit the offence under s. 87(1) of driving otherwise than in accordance with a licence (**see chapter 11**).

A person may not be granted such a licence unless the Secretary of State is satisfied that he/she is, in relation to his/her conduct, a fit person to hold such a licence (s. 112). A person may appeal against any refusal under s. 119.

'Conduct' has to be considered in context and in the round. Therefore a conviction of a driver for indecent assault on a girl passenger (resulting in his being required to register as a sex offender) was a relevant consideration but was by no means determinative of the issue without consideration of the other circumstances of his private and working life (*Secretary of State for Transport, Local Government and the Regions* v *Snowdon (2002) LTL 4 November*).

Part IV of the 1988 Act provides for the revocation of these licences and the disqualification of holders.

Application for an LGV or PCV licence is made under the Motor Vehicles (Driving Licences) Regulations 1999 (SI 1999/2864), part IV. Those Regulations set out the requirements for the relevant tests which much be passed, together with the requirements of provisional licences and learner drivers. Note that holders of passenger-carrying vehicle provisional licences must not carry passengers other than a supervising qualified driver (as to which, **see chapter 11**) or a full licence holder who is giving/receiving instruction (reg. 16(8)). These restrictions do not apply during a driving test.

13.2.4 Disqualification

Where a person is disqualified by a court (**see chapter 11**) from driving, that person may also have his/her LGV/PCV licence revoked for such a period as the licensing authority thinks fit (reg. 56 of the 1999 Regulations).

If the holder of an LGV licence is under 21 years of age and he/she accumulates more than three penalty points (or is disqualified) that person *must* be disqualified from holding an LGV licence:

- indefinitely *or*
- at least until reaching the age of 21

(s. 115(1)(a) of the Road Traffic Act 1988 and reg. 55 of the 1999 Regulations).

Large goods vehicles and passenger carrying vehicles licences last until the drivers' 45th birthday or five years, whichever is the longer. If the driver is aged between 45 and 65 they last for five years or until the 66th birthday, whichever is the *shorter*. After the driver has reached 65 the licence must be renewed annually (s. 99(1A)(a)). (Contrast ordinary licences; **see chapter 11**.)

13.2.5 Exemptions

Regulation 51 of the 1999 Regulations exempts certain classes of vehicle which would otherwise require LGV or PCV licences. These include some agricultural, forestry and road-rolling vehicles, together with steam-driven vehicles and industrial plant.

If a small PCV (**see para. 13.2.2**) is driven for social purposes on behalf of a non-commercial body, other than hire or reward, there is no need for a PCV licence. The driver must however be over the age of 21 and have held an ordinary licence for at least two years (reg. 7).

Regulation 7 imposes further conditions on the driver if he/she is aged 70 or over and makes additional restrictions on the weight of any small PCV that can be so driven.

Regulation 7 also makes provision for certain people to drive passenger carrying vehicles for the purposes of repair or testing.

Police

Regulation 50 provides a general exemption from the requirements of Part IV in respect of passenger-carrying vehicles being driven by a constable for the purposes of:

- removing or avoiding obstruction
- to other road users or other members of the public, *or*
- protecting life or property (including the vehicle and its passengers), *or*
- 'other similar purposes'.

There is no requirement that the constable be in uniform. For obstruction generally, **see chapter 8**.

13.3 Licences

13.3.1 Operators' Licences

In addition to the need for drivers to hold certain specialised driving licences, there are also requirements for those who wish to *use* goods vehicles in the course of their business.

The use of goods vehicles for business purposes is governed by the Goods Vehicles (Licensing of Operators) Act 1995 and the similarly-named Regulations (SI 1995/2869).

If a person wishes to use a goods vehicle on a road for hire or reward or in connection with his/her trade or business, he/she will need an 'operators' licence' (s. 2).

Such a licence is very different from a driving licence and amounts to an authority to operate a number of goods vehicles.

In line with an EC Directive (96/26/EC, 23 May 1996), there are two types of operators' licence; standard and restricted.

Standard Operators' Licence

Vehicles under this licence may be used for:

- hire or reward *or*
- in connection with the holder's trade or business.

Standard operators' licences are divided into those which allow the above activities to be carried out internationally and those which may only be used nationally.

Restricted Operators' Licence

Vehicles under this licence may be used in connection with the holder's trade or business *other than the carriage of goods for hire or reward.*

Whether or not a goods vehicle is being used for hire or reward will be a matter of fact for a court to determine (for guidance, see *Albert* v *Motor Insurers' Bureau* [1971] 2 All ER 1345).

Given the much wider effect of the standard operators' licence, holders of them are subject to much more stringent control than those with restricted licences.

Operators' Licence Generally

Under the 1995 Regulations, applicants for operators' licences have to demonstrate that they are of good repute, that they are professionally competent and that they have the appropriate financial standing. These requirements have been tightened up by the Goods Vehicle Operators (Qualifications) Regulations 1999 (SI 1999 No. 2430), particularly in relation to the applicant's record of road traffic offences.

In determining these matters, and the fitness of applicants for LGV/PCV licences, the traffic commissioners have a significant role to play (see s. 111 of the Road Traffic Act 1988). However, in deciding the fitness or otherwise of applicants and operators, the Commissioners are bound by the general principles of proportionality under the European Convention on Human Rights and must consider the effect of their decision on the person's Convention rights (such as the right to peaceful enjoyment of property—**see General Police Duties, chapter 2**) (*Crompton* (*T/A David Crompton Haulage*) v *Department of Transport North Western Area* (2003) LTL 31 January).

The 1995 Regulations, together with the provisions of the 1995 Act itself, set out requirements for:

- the specification of vehicles which an operator can use
- the operating centre at which the vehicles are kept and despatched
- environmental controls on operating centres
- conditions attached to the operators' licence.

Regulation 23 requires a disc to be displayed on the vehicle stating, among other things, what type of licence the holder has. In the case of a standard (national) licence, the disc is blue, for a standard (international) licence the disc is green and a restricted licence the disc is orange.

OFFENCE: **Using Vehicle without Operators' Licence—*Goods Vehicles (Licensing of Operators) Act 1995, s. 2***

- Triable summarily • Fine

(No specific power of arrest)

The Goods Vehicles (Licensing of Operators) Act 1995, s. 2 states:

(1) Subject to subsection (2) and section 4, no person shall use a goods vehicle on a road for the carriage of goods—
  (a) for hire or reward, or
  (b) for or in connection with any trade or business carried on by him,
  except under a licence issued under this Act; and in this Act such a licence is referred to as an 'operator's licence'.

(2) Subsection (1) does not apply to—
  (a) the use of a small goods vehicle within the meaning given in Schedule 1;
  (b) the use of a goods vehicle for international carriage by a haulier established in a member State other than the United Kingdom and not established in the United Kingdom;
  (c) the use of a goods vehicle for international carriage by a haulier established in Northern Ireland and not established in Great Britain; or
  (d) the use of a vehicle of any class specified in regulations.

(3) ...

(4) It is hereby declared that, for the purposes of this Act the performance by a local or public authority of their functions constitutes the carrying on of a business.

KEYNOTE

For an explanation of 'using' a vehicle, see chapter 1.

It would appear that it is necessary to prove that the vehicle was actually carrying goods at the relevant time (see *Robertson* v *Crew* [1977] RTR 141).

The duty to furnish information in relation to the identity of the driver at the time of an alleged offence under this section is to be found in the Road Traffic Act 1960 (s. 232).

For enforcement powers in relation to offences under s. 2, see para. 13.3.3.

OFFENCE: **Contravening Condition of Operators' Licence—*Goods Vehicles (Licensing of Operators) Act 1995, s. 22(6)***

- Triable summarily • Fine

(No specific power of arrest)

The Goods Vehicles (Licensing of Operators) Act 1995, s. 22 states:

(6) Any person who contravenes any condition attached under this section to a licence of which he is the holder is guilty of an offence...

KEYNOTE

One of the conditions which must be attached to the licence is a duty to report any changes to the traffic commissioners. Such changes will usually include any convictions of the holder or the transport manager or any changes in their financial standing, good repute, etc. above. Such notification must be made within 28 days of the event (or of the event coming to the knowledge of a transport manager).

13.3.2 Exemptions

Section 2(2) of the 1995 Act sets out exemptions for 'small' goods vehicles and for vehicles being used by a haulier established in Northern Ireland or a European Union member State.

Small goods vehicle includes a vehicle with a 'plated' weight of 3.5 tonnes or less or, if there is no plated weight, then an unladen weight of 1,525 kgs. There are further calculations to be made if trailers are involved and sch. 1 to the 1995 Act should be consulted.

Other exempted vehicles include emergency services vehicles, vehicles being used for the provision of emergency supplies by utilities companies, funeral vehicles and 'showman's vehicles'.

Vehicles temporarily in Great Britain are covered by the obviously-named Goods Vehicles (Licensing of Operators) (Temporary Use in Great Britain) Regulations 1996 (SI 1996/2186). The Regulations are extremely detailed and should be consulted in full.

If a vehicle is brought into the country frequently enough, it may cease to be 'temporary' (*BRS* v *Wurzal* [1971] 3 All ER 480).

13.3.3 Enforcement

The Goods Vehicles (Licensing of Operators) Act 1995 enables regulations for enforcement of some of its provisions to be made (see sch. 1A). The Goods Vehicles (Enforcement Powers) Regulations 2001 (SI 2001/3981) are such and they came into force in January 2002.

In summary, the regulations empower an authorised person to detain a vehicle and its contents where the person:

- has reason to believe
- that the vehicle is being or
- has been used
- on a road

in contravention of s. 2 of the Act (reg. 3). However, the regulation does not authorise anyone other than a constable in uniform to stop the vehicle in order to detain it and its contents (reg. 3(2)).

When a vehicle has been detained under reg. 3, there is the very quaint requirement to put a notice to that effect in the *London Gazette* (reg. 9). On a more practical note, reg. 9 also requires a copy of that notice to be served on a number of people including the owner of the vehicle and the chief officer of police and traffic commissioner for the area in which the vehicle was detained. The owner of a vehicle detained under reg. 3 (but not, it seems, the owner of the *contents* alone) can apply to a traffic commissioner for its return based on one of a number of conditions (generally arguing lawful use or lack of knowledge by the owner—see reg. 10). The traffic commissioner may hold a hearing to determine the application; if the owner is not happy with the response, they can appeal to a Transport Tribunal (reg. 13).

Regulation 4 provides for circumstances where the vehicle must be released (basically when the person using the vehicle has a licence or if there is no offence under s. 2).

Under reg. 5 an authorised person may attach an immobilisation device to the vehicle where it is found or they can move the vehicle or require it to be moved to a more convenient place and attach an immobilisation device there. Any immobilisation device must be accompanied by an appropriate notice saying that the device cannot be removed without authorisation and that no attempt should be made to drive the vehicle or otherwise put it in motion.

OFFENCE: **Interfering with Immobilisation Device or Notice and Obstruction—*Goods Vehicles (Enforcement Powers) Regulations 2001, regs 6 and 20***

- Triable summarily • Fine

(No specific power of arrest)

The Goods Vehicles (Enforcement Powers) Regulations 2001 state:

> 6.—(1) An immobilisation notice shall not be removed or interfered with except by or on the authority of an authorised person.
>
> (2) A person contravening paragraph (1) shall be guilty of an offence . . .

(3) Any person who, without being authorised to do so in accordance with regulation 7 removes or attempts to remove an immobilisation device fixed to a vehicle in accordance with these Regulations shall be guilty of an offence...

20. —Where a person intentionally obstructs an authorised person in the exercise of his powers under regulation 3 or 8, he is guilty of an offence...

KEYNOTE

The proper procedure for releasing a vehicle immobilised under reg. 6 is set out in reg. 7; this requires that it be released under the direction of an authorised person. As with detention of the vehicle and its contents generally, if it turns out that the person using the vehicle had a licence or if there is no offence under s. 2, the immobiliser must be removed (reg. 7(b)).

An authorised person can give written direction that any property detained under reg. 3 be delivered to a specified 'appointed' person and also that a vehicle be driven, towed or removed using reasonable means (reg. 8). The regulations contain further detail about the types of people who can receive detained property and vehicles and also about the breaking of journeys to drop off a vehicle's contents.

13.4 Public Service Vehicles

The operation of public service vehicles (PSVs) is contained in the Public Passenger Vehicles Act 1981, as amended, and the Transport Act 1985.

The legislation is very detailed and what follows is a brief outline of some of the relevant features.

Like the operation of goods vehicles, the use of PSVs is closely controlled although one aim of the Transport Act 1985 was to deregulate controls on local services.

Under s. 1 of the Public Passenger Vehicles Act 1981 a public service vehicle is:

(1) a motor vehicle (other than a tramcar) which—
 (a) being a vehicle adapted to carry more than eight passengers, is used for carrying passengers for hire or reward; or
 (b) being a vehicle not so adapted, is used for carrying passengers for hire or reward at separate fares in the course of a business of carrying passengers.

KEYNOTE

The use that is made of the particular PSV and the payment by passengers is important in determining whether the legislation applies or not:

- 'Stage carriages' are PSVs being used for local services, that is, used for carrying passengers for hire or reward on specified short journeys.
- 'Express carriages' are PSVs being used for carrying passengers for hire or reward at separate fares, other than on local services.
- 'Contract carriages' are PSVs being used for carrying passengers for hire or reward other than at separate fares.

Local or 'stage' services differ from 'express' services in the length of journey being made. If the passengers are picked up and dropped off within a relatively short distance (as with a traditional bus service), each paying the relevant fare, the service is likely to be a 'stage' service using 'stage carriages'.

If the passengers are taken 15 miles or more (e.g. from one part of the country to another), the service is likely to be an 'express' service and the PSV an 'express' carriage.

'Hire or reward' does not only mean a service provided under some contractual arrangement. It will include a provision of a transport facility amounting to a business activity that goes beyond simple kindness (see *DPP v Sikondar* [1993] RTR 90).

Hotel courtesy buses have been held to be PSVs on the basis that the service was part of the overall hotel 'service', the cost of which was included in the price of a room or meal (*Rout* v *Swallow Hotels Ltd* [1993] RTR 80).

If a person hires an entire PSV for a particular journey or journeys (e.g. a club outing or a 'hen' or 'stag' night), this is a 'contract' service using a 'contract carriage'.

Local or stage services must be registered with the relevant traffic commissioners (in London a special local service licence is required).

Provision of local services in London is governed by Part V of the Greater London Authority Act 1999. The 1999 Act creates the 'London bus network' and imposes additional requirements upon people providing London local services that fall outside the network.

13.4.1 Operators' Licences

All PSVs being used on a road for hire or reward require an operators' licence, irrespective of the type of service being provided (s. 12(1) of the Public Passenger Vehicles Act 1981).

Operators' licences are obtained from the relevant traffic commissioners and are categorised in the same way as goods vehicles operators' licences, namely standard and restricted licences.

Again, operators must show that they are of good repute and financial standing in order to be granted an operators' licence. In the case of a standard licence, they must also demonstrate professional competence. These requirements have been tightened up under the Public Service Vehicle Operators (Qualifications) Regulations 1999 (SI 1999/2431).

An operators' disc must be exhibited on the relevant vehicle(s) and the Public Service Vehicles (Operators' Licences) Regulations 1995 set out other conditions which must be followed by the holder of a licence.

Traffic commissioners are required to attach certain conditions to an operators' licence in relation to the numbers of vehicles covered by that licence; they may also attach other conditions as they think fit (s. 16 of the Public Passenger Vehicles Act 1981). One such condition (under s. 16A) is the requirement for a licence holder to inform the traffic commissioners of any event that might affect the licence holder's good repute, competence, etc. This condition was added to the 1981 Act by the 1999 Regulations above.

Conditions may also be attached under ss. 16, 26 and 27 of the Transport Act 1985.

Contravention of a condition of an operators' licence is an offence under s. 16(7) of the Public Passenger Vehicles Act 1981.

Standard PSV Operators' Licence

This licence authorises the use of any PSV for both national and international journeys. Note that there is now a requirement for anyone carrying passengers in public service vehicles travelling between member States of the EU to hold a Community licence (see the Public Service Vehicles (Community Licence) Regulations 1999 (SI 1999/1322). A copy of the licence must be kept on each public service vehicle carrying out an international transport operation and the original or a certified true copy must be produced on demand to an 'authorised' inspecting officer which includes a police constable (reg. 8).

Restricted PSV Operators' Licence

This licence authorises the use of certain smaller PSVs for both national and international journeys. Taxi licence-holders (**see para. 13.7**) may apply for a restricted licence under s. 12 of the Transport Act 1985 (referred to as a 'special licence').

13.4.2 Exemptions

There are many exemptions to the requirements for an operators' licence. These include education authorities using school buses, small PSVs being used privately and other vehicles being used for charitable purposes.

Some foreign PSVs are also exempt from these provisions.

13.4.3 Conduct of People on PSVs

The duties and powers of drivers and conductors on PSVs is regulated by statutory instrument.

Most of the conditions are contained in the Public Service Vehicles (Conduct of Drivers, Inspectors, Conductors and Passengers) Regulations 1990 (SI 1990/1020), as amended.

Regulations are made under the two Acts referred to above. The Regulations address the conduct of bus crews generally and impose certain duties upon them (e.g. taking all reasonable precautions to ensure passenger safety).

Regulations 6 to 7 also make requirements of passengers in relation to the payment of fares, altering tickets, and general anti-social or dangerous behaviour (including the playing of things such as personal stereos to the annoyance of other passengers—reg. 6(1)).

Clearly, in more serious cases, offences under the Public Order Act 1986 or the Protection from Harassment Act 1997 might be considered (**see General Police Duties, chapter 3**).

Under regs 8 and 9 passengers can be required to give their name and address to a driver, inspector or conductor if those people reasonably suspect the passenger to have contravened one of the regulations.

13.4.4 Disabled Access

The Public Service Vehicles Accessibility Regulations 2000 (SI 2000/1970) introduced many technical modifications that must be made to certain types of PSV to ensure that they are readily accessible to people with physical disabilities. The 2000 Regulations were made under the Disability Discrimination Act 1995 and are primarily designed to ensure that people with disabilities are able to get on and off 'regulated' PSVs safely and without unreasonable difficulty and that people using wheelchairs are able to remain in those wheelchairs when using these PSVs. 'Regulated' PSVs are generally coaches and buses (both single and double-decked) having a capacity for more than 22 passengers and which are used to provide local and scheduled services. The exact technical requirements are set out in the three schedules to the 2000 Regulations but, very broadly, they make requirements for boarding lifts/ramps, entrances and exits, communication devices (e.g. boarding aid alarms to indicate that equipment such as lifts are in operation), handrails and lighting.

Further changes to the requirements for PSVs' equipment in this regard can be found in the Public Service Vehicles (Conditions of Fitness, Equipment, Use and Certification) (Amendment) Regulations 2002 (SI 2002/335).

13.5 Community Buses and Trams

13.5.1 Community Buses

Sections 22 and 23 of the Transport Act 1985 make provision for the operation of 'Community bus services' which are services provided for the social and welfare needs of one or more communities.

Drivers of 'community buses' may hold a community bus permit which exempts them from the need for a PCV licence and the conditions relating to such a permit can be found in the Community Bus Regulations 1978 (SI 1978/1313) and 1986 (SI 1986/1254) (as amended in each case).

Under the 1978 Regulations, community buses must display appropriate discs on the inside of the vehicle but where they can be read easily from outside the vehicle.

Community bus services are subject to the conditions laid down in the Transport Act 1985, which also creates a number of offences.

13.5.2 Tramcars

Under s. 193A of the Road Traffic Act 1988, the Secretary of State may make amendments to various Acts so that they include tram and trolley vehicles.

Under this power the Tramcars and Trolley Vehicles (Modification of Enactments) Regulations 1992 (SI 1992/1217) have been made. Many local variations exist in the regulation of tram and trolley services (e.g. the Stage Carriages Act 1832 still applies to trams in Blackpool), and reference should be made to those local regulations in each case.

In addition, the Transport and Works Act 1992 provides for certain offences (including drink/driving (see Part II of the Act)) committed by people employed in guided transport systems (which also include railways)—**see General Police Duties, chapter 13**.

13.6 Drivers' Hours

Drivers of goods vehicles or passenger carrying vehicles have, in the past, come under considerable commercial pressure to drive for long distances or for long hours. In order to reduce the increased risk which is presented to the public by drivers of those vehicles who are suffering from fatigue, the Transport Act 1968 (Part VI) introduced a system for the recording and monitoring of the hours which some drivers work.

This was achieved by setting out maximum periods of driving, together with minimum rest periods in between and by requiring certain vehicles to have tachographs (**see paras 13.6.3 and 13.6.4**) fitted and maintained.

The law has been extended—and considerably complicated—by the effects of EC legislation and is now a highly complex area of road traffic law. The many 'Community rules' (see s. 96 of the Transport Act 1968) and exemptions should be consulted, and expert assistance sought in each case.

Although these rules are strictly enforced and breaches are viewed very seriously by the courts (see below), they are occasionally relaxed temporarily as happened during the fuel shortage during 2000 when tanker drivers were given temporary exemption from the relevant legislation (see the Community Drivers' Hours (Passenger and Goods Vehicles) (Temporary Exception) Regulations 2000 (SI 2000/2483) and more recently in connection with the foot and mouth outbreak).

What follows here is a *very brief overview* of the legal regulation of drivers' hours.

13.6.1 Type of Vehicle

The provisions of Part VI of the Transport Act 1968 apply generally to:

- **Passenger vehicles**—PSVs and other motor vehicles constructed or adapted to carry more than 12 passengers.
- **Goods vehicles**—locomotives, motor tractors, any motor vehicle constructed so that it may be used as an articulated vehicle and other motor vehicles constructed or adapted to carry goods other than passengers' effects.

As the purpose of the legislation is to exercise control over working hours a number of exemptions are provided for. These essentially refer to vehicles used for the conveyance of passengers for short distances, public service and public utility vehicles, ambulances etc., types of tractors, showmens' goods vehicles and some breakdown vehicles. Most of the exempted vehicles are controlled by public authorities or public utilities.

The rules and restrictions vary according to the type of *vehicle*; they also vary according to the type of *journey* being made by that vehicle.

13.6.2 Type of Journey

The relevant types of journey or work are:

- international journeys
- national journeys
- domestic journeys
- mixed (i.e. a combination of some of the above).

As ever, there are many exemptions to the requirements for observing and recording drivers' hours and these generally appear in the particular rules.

Community Legislation

Most journeys will be covered by:

- Regulation 85/3820 (in relation to drivers' hours); and
- Regulation 85/3821, as amended (in relation to tachograph records).

13.6.3 Drivers' Hours

The following is a brief résumé of the regulations in relation to drivers' hours.

Driving Periods

- Normally 9 hours daily driving period but 10 hours not more than twice a week.
- Generally 4.5 hours continuous driving period.
- The total period of driving in any one week must not exceed 56 hours.
- The total period of driving in any one fortnight must not exceed 90 hours.

These figures are an aggregation of the various prescribed driving and rest periods found within the relevant articles.

Rest Breaks

- At least 45 minutes rest break unless the driver is beginning a daily or weekly rest period.
- Rest breaks may be replaced by rest periods of at least 15 minutes each totalling 45 minutes distributed within the continuous driving period or immediately after it.
- Rest breaks must not be taken as part of a daily rest period.

Rest Periods

- After 6 daily driving periods drivers must take a weekly rest period.
- The weekly rest period may be postponed until the end of the sixth day if the total driving time over the 6 days does not exceed the maximum corresponding to 6 daily driving periods.

- In the case of national and international carriage of passengers other than on regular services (i.e. coach tours) it is 12 and not 6 daily driving periods.
- Drivers must have a daily rest period of at least 11 consecutive hours in each period of 24 hours.
- The rest period may be reduced to at least 9 consecutive hours not more than three times a week. This applies provided that an equivalent period is granted as compensation before the end of the following week.

The Divisional Court has held that employers are under a duty to conduct periodic checks of tachograph charts even though there is no reason to suspect any breach (*Vehicle Inspectorate* v *Nuttall (t/a Redline Coaches)*[1998] RTR 321).

13.6.4 Offences

Section 97 of the Transport Act 1968 states:

> (1) It is an offence for a person to use, or cause or permit to be used, a vehicle to which this section applies unless there is recording equipment in the vehicle which complies with the Community Recording Equipment Regulation.

Section 97A of the Transport Act 1968 states:

> It is an offence for an employed driver to fail without reasonable excuse to return any record sheet which relates to him to his employer within 21 days of completing it; or where he has two or more employers by whom he is employed as a driver of such a vehicle, to notify each of them of the name and address of the other or others of them.

Section 97AA of the Transport Act 1968 states:

> (1) It is an offence for a person, with intent to deceive, to forge, alter or use any seal on recording equipment installed in or designed for installation in, a vehicle to which s. 97 applies.

Section 98 of the Transport Act 1968 states:

> (4) It is an offence for any person to contravene any regulations made or any requirements as to books, records or documents.

Section 99 of the Transport Act 1968 states:

> (4) It is an offence for any person to fail to comply with any requirement to allow the inspection of records or to obstruct an officer in the exercise of his powers under s. 99.

KEYNOTE

In answer to a charge of tampering or interfering with tachograph equipment, there is a statutory defence (under s. 97(4)) where the breaking or removal of the machine's seal could not have been avoided and that it had not been reasonably practicable for the seal to be replaced in an authorised manner. As this offence is one of 'strict liability' (as to which, **see Crime, chapter 1**) this defence is a very narrow one (see *Vehicle Inspectorate* v *Sam Anderson (Newhouse) Ltd* [2002] RTR 217).

The seriousness with which the courts view such offences as these was illustrated in a recent case involving the deliberate alteration of tachograph records by a number of drivers. The drivers received sentences of eight months' imprisonment and these sentences were upheld by the Court of Appeal (*R* v *Saunders* [2002] RTR 4).

13.6.5 Police Powers

Under s. 99(1) of the Transport Act 1968, a constable may require any person to produce and permit him/her to inspect and copy any book or register which the person is required by regulations under s. 98 of the 1968 Act to carry or have in his/her possession for the purpose of making in it:

- any entry required by those regulations or which is required to be carried on any vehicle of which that person is the driver;
- any book or register which that person is required to preserve;
- any record sheet which that person is required by Article 14(2) of the Community Recording Equipment Regulation (1985/3821) to retain or to be able to produce (Article 15(7) of the Regulation);
- if that person is the owner of a vehicle to which the Transport Act 1968 applies, any other document of that person which the officer may reasonably require to inspect for the purpose of ascertaining whether the provisions of the Act, or of regulations made thereunder, have been complied with;
- any book, register or document required by the applicable Community rules or which the officer may reasonably require to inspect for the purpose of ascertaining whether the requirements of the applicable Community rules have been complied with;
- and that record sheet, book, register or document shall, if the officer so requires by notice in writing served on that person, be produced at the office of the traffic commissioner specified not less than 10 days from the service of the notice.

For the purpose of exercising these powers an officer may detain the vehicle in question during such time as is required for the exercise of that power.

The House of Lords has held that the requirement to 'produce' records and 'permit' inspection includes the handing over of those records; it also includes allowing the retention of those records for the purposes of inspection and the taking away of the records for effective and thorough examination if appropriate (*Cantabrica Coach Holdings Ltd* v *Vehicle Inspectorate* [2001] 1 WLR 2288). Their Lordships also held that it was a matter for the discretion of the 'authorised officer' to decide whether to go to the offices of the relevant operator and require production of those records, or to give ten days' notice requiring the records to be produced at the offices of the traffic commissioner.

These powers are also available to examiners authorised under s. 66A or part IV of the Road Traffic Act 1988 or authorised by the traffic commissioners (s. 99(2)), provided that, in each case, the examiner produces the requisite authority.

If the driver of a UK registered vehicle obstructs an authorised person in the exercise of the above powers or fails to comply with a requirement made under s. 99(1) above, the authorised person may prohibit the driving of the vehicle on a road (s. 99A). Where such a prohibition is made, the authorised person may also direct the driver to remove the vehicle (and any trailer it is drawing) to a place specified by him/her. Driving a vehicle, or causing or permitting a vehicle to be driven in contravention of such an order is a summary offence punishable with a fine. Similarly, refusing or failing to comply with a direction made under s. 99A (within a reasonable time) is a summary offence punishable by a fine (s. 99C).

There is a requirement for a person 'keeping' a vehicle involved in an alleged offence under s. 99 to provide details leading to the identification of the driver when required to do so by police (Road Traffic Act 1960, s. 232).

13.6.6 AETR Agreement

Foreign vehicles operating in the UK and UK vehicles making certain overseas journeys may also be subject to the provisions of the European Agreement concerning the Work of Crews of Vehicles engaged in International Road Transport, known as the AETR Agreement (Cmnd 7401).

The contents of the AETR agreement are outside the scope of this Manual.

13.6.7 Vehicles Carrying Loads of Exceptional Dimensions

Under s. 44(1)(d) of the Road Traffic Act 1988, the Secretary of State may authorise the use on roads of vehicles or trailers carrying loads of exceptional dimensions that would otherwise be unlawful under various road traffic law provisions. Such vehicles usually attract higher excise duty. Exceptional loads are, in broad terms, loads that are too big to be carried by a 'heavy motor car' (as to which, **see chapter 1**) with or without a trailer, or too heavy to be carried by such a motor car with a laden weight of not more than 41,000 kg (see the Vehicle Excise and Registration Act 1994, sch. 1, para. 6). The importance of such loads and their movement is that Community Support Officer designated under sch. 4 to the Police Reform Act 2002 (**see General Police Duties, chapter 2**) and employees accredited under sch. 5 to the Act may be empowered to direct traffic in connection with the escorting of a vehicle or trailer carrying such loads.

13.7 Taxis

The law regulating the use of taxis and hackney carriages is mainly provided for in four statutes:

- the Town Police Clauses Act 1847
- the Local Government (Miscellaneous Provisions) Act 1976
- the Metropolitan Public Carriage Act 1869
- the Private Hire Vehicles (London) Act 1998.

Under the Greater London Authority Act 1999 responsibility for regulating taxis and minicabs in London is transferred from the police and the Secretary of State to Transport for London (as to which, **see chapter 10**), and the Private Hire Vehicles (London) Act 1998 will be amended accordingly.

Taxis—vehicles which ply for hire—are generally referred to in the relevant legislation as hackney carriages; those which do not ply for hire are referred to as private hire vehicles. For the relevant regulations in London, see the Greater London Authority Act 1999, schs 20 (hackney carriages—some of which was not in force at the time of writing) and 21 (private hire vehicles).

What follows is a brief review of some of the relevant legislation in this area.

13.7.1 Licensing System

Under the Town Police Clauses Act 1847 local authorities are empowered to provide a licensing system for hackney carriages to ply for hire as taxis (s. 37). In England and Wales that licensing system is to apply throughout the whole of a district council area (s. 15 of the Transport Act 1985).

References to licences in Part II of the Local Government (Miscellaneous Provisions) Act 1976 are to licences issued by a particular district. The effect of this is that each operator, driver and vehicle will require a valid licence in respect of each district area in which they work.

Under s. 16 the council can limit the number of hackney carriage licences it issues, provided that it is satisfied that there is no significant demand for those services (see *Ghafoor* v *Wakefield Metropolitan District Council* [1990] RTR 389).

In London the system for hackney carriages and private hire vehicles has been provided under the Metropolitan Public Carriage Act 1869. From 22 January 2001, application can be made for a London Private Hire Vehicle Operators' licence under the Private Hire Vehicles

(London) Act 1998. The licensing authority for this scheme is Transport for London. The 1998 Act creates a number of offences connected with inviting or accepting of private hire bookings by people who do not hold a London PHV operators' licence. More specific detail regulating the granting of such licences, together with the records that must be kept by licence holders can be found in the Private Hire Vehicles (London) (Operators' Licences) Regulations 2000 (SI 2000 No. 3146).

13.7.2 Application

An applicant for a licence must satisfy the granting authority that he/she is a fit and proper person. The authority may decide against the application on the civil law standard of a 'balance of probabilities', even where it is considering the applicant's acquittal for a criminal offence (see *R* v *Maidstone Crown Court, ex parte Olson* (1992) *The Times*, 21 May).

In arriving at its decision, the authority may seek the advice of the local police (see s. 47 of the Road Traffic Act 1991).

The main relevant licences are:

- Taxi drivers' licences granted under the Town Police Clauses Act 1847 and the Metropolitan Public Carriage Act 1869.
- Private hire drivers' licences granted under s. 51 of the Local Government (Miscellaneous Provisions) Act 1976.
- Private hire proprietors' licences granted under s. 48 of the 1976 Act.
- Private hire operators' licences granted under s. 55 of the 1976 Act.

There are summary offences of operating, driving or plying for hire without the relevant licence. The Divisional Court has held that 'plying for hire' can extend to a situation where the vehicle is in a prominent position on private property just off a street and the public were on the street in numbers (*Eastbourne Borough Council* v *Stirling* [2001] RTR 65).

It is also a summary offence to tout for hire car services in a public place (see s. 167 of the Criminal Justice and Public Order Act 1994). This offence is an 'arrestable' offence under s. 24(2) of the Police and Criminal Evidence Act 1984 (as to which, **see General Police Duties, chapter 2**).

Taxi and private hire licences (under the Town Police Clauses Act 1847 and the Local Government (Miscellaneous Provisions) Act 1976) remain in force for any time up to three years and must be produced on request to a police officer or authorised officer of the council (s. 53(3) of the Local Government (Miscellaneous Provisions) Act 1976). Failure to do so will not be an offence if it is produced within seven days.

A council may attach such conditions as it considers reasonably necessary to the licences it issues, including the markings and appearance of the vehicle (s. 47 of the Local Government (Miscellaneous Provisions) Act 1976). However, each of the many licensing provisions is open to an appeal process to a magistrates' court (see *R* v *Blackpool Borough Council, ex parte Red Cab Taxis Ltd* [1994] RTR 402).

If a vehicle is licensed as a hackney carriage in one district, it can also be licensed as a private hire vehicle in another district (*Kingston-upon-Hull City Council* v *Wilson* (1995) *The Times*, 25 July).

If a licensed hackney carriage gives the appearance of such it cannot be used without a current licence, even in a private capacity as a family motor car by the owner (*Darlington Borough Council* v *Thain*, 23 November 1994, unreported).

Licence plates are issued under s. 35 of the Local Government (Miscellaneous Provisions) Act 1976 and remain the property of the issuing council. Using or permitting the use of a private hire vehicle without displaying such a plate is a summary offence (s. 48).

Note that a local authority may bring a prosecution against taxi drivers under s. 143 of the Road Traffic Act 1988 where they have no insurance (*Middlesbrough Borough Council* v *Safeer* [2002] 1 Cr App R 266). For the offence of having no insurance, **see chapter 6**.

13.7.3 Exemptions

The legislation allows for a number of exemptions to the licensing provisions (see s. 75 of the 1976 Act). These include the use of hire vehicles for funerals and weddings and for contractual hire periods of at least seven days. In such cases, the onus is on the defendant to establish that he/she falls within one of the exemptions (*Leeds City Council* v *Azam and Fazi* [1989] RTR 66).

14 Fixed penalty system

14.1 Introduction

Originally introduced by the Transport Act 1982 the fixed penalty system is intended to speed up the administrative procedure for the prosecution of certain road traffic offences. The process is initiated by serving a person with a fixed penalty notice, attaching such a notice to a stationary vehicle or through the 'conditional offer' system.

The main elements of the first two of these processes are set out briefly below. However reference should be made to the legislation itself for a full explanation of the procedure to be followed in each case. The 'conditional offer' is dealt with at the end of this chapter.

The system described in this chapter has no connection with the system for dealing with other instances of criminal conduct by way of fixed penalties under the Criminal Justice and Police Act 2001 (as to which, **see General Police Duties**).

14.2 Extended Fixed Penalty System

The law governing the extended fixed penalty system is to be found under Part III of the Road Traffic Offenders Act 1988.

The 'fixed penalty' itself is determined by s. 53 which allows for the setting of the level of any penalty to be imposed under the system.

The purpose of the system is to avoid court hearings for the listed offences but, at the same time, to ensure that those committing them receive the appropriate punishment including the endorsement of their licence. The definition of a 'fixed penalty notice', under s. 52 sums up the essence of the system:

> (1) 'A 'fixed penalty notice' means a notice offering the opportunity of the discharge of any liability to conviction of the offence to which the notice relates by payment of a fixed penalty...

Where a fixed penalty offence (see below) has been committed, the system allows for a procedure other than prosecution to be followed during a specified period. This period, known as the 'suspended enforcement period' (s. 52(3)(a)), is at least 21 days following the date of the fixed penalty notice or such longer period as may be specified in it.

14.3 Fixed Penalty Offences

The offences to which the system applies are set out in sch. 3 to the Road Traffic Offenders Act 1988 (**see appendix 3**).

Section 51(2) of the 1988 Act, however, provides that an offence specified in the schedule is not a fixed penalty offence if it is committed by *causing* or *permitting* a vehicle to be used by another in contravention of any statutory provision, restriction or prohibition.

Effectively this means that defendants reported for causing or permitting offences may not enjoy the administrative provisions made under the fixed penalty system.

Further offences may be added or removed from the system by s. 51(3). There are regular Fixed Penalty Orders to achieve this so the schedule is subject to change.

The Road Traffic (Vehicle Emissions) (Fixed Penalty) Regulations 1997 (SI 1997/3058), provide for certain specified local authorities to issue fixed penalty notices to vehicle users who contravene regs 61 and 98 of the Road Vehicles (Construction and Use) Regulations 1986 (**see chapter 9**) in relation to the emission of smoke and fumes.

14.4 Fixed Penalty Procedure

As mentioned above, the fixed penalty procedure envisages two situations; where the driver is present and where there is a stationary vehicle to which a notice can be fixed. Where the relevant offence is endorsable, the procedure makes provision for the submission by a driver of his/her licence. This is to allow the clerk at the relevant court to record the endorsement on the licence.

The procedure in each case has been briefly summarised in **appendix 9, diagrams 1 to 3**.

Proof of the service of fixed penalty notices by police officers is provided for under s. 79 of the 1988 Act.

14.4.1 Procedure Where Driver is Present

Section 54 of the Road Traffic Offenders Act 1988 states:

(1) This section applies where in England and Wales on any occasion a constable in uniform has reason to believe that a person he finds is committing or has on that occasion committed a fixed penalty offence.

(2) Subject to subsection (3) below, the constable may give him a fixed penalty notice in respect of the offence.

(3) Where the offence appears to the constable to involve obligatory endorsement, the constable may only give him a fixed penalty notice under subsection (2) above in respect of the offence if—
   (a) he produces his licence and its counterpart for inspection by the constable,
   (b) the constable is satisfied, on inspecting the licence and its counterpart, that he would not be liable to be disqualified under section 35 of this Act if he were convicted of that offence, and
   (c) he surrenders his licence and its counterpart to the constable to be retained and dealt with in accordance with this Part of this Act.

(4) Where—
   (a) the offence appears to the constable to involve obligatory endorsement, and
   (b) the person concerned does not produce his licence and its counterpart for inspection by the constable,

   the constable may give him a notice stating that if, within seven days after the notice is given, he produces the notice together with his licence and its counterpart in person to a constable or authorised person at the police station specified in the notice (being a police station chosen by the person concerned) and the requirements of subsection (5)(a) and (b) below are met he will then be given a fixed penalty notice in respect of the offence.

(5) If a person to whom a notice has been given under subsection (4) above produces the notice together with his licence and its counterpart in person to a constable or authorised person at the police station specified in the notice within seven days after the notice was so given to him and the following requirements are met, that is—
   (a) the constable or authorised person is satisfied, on inspecting the licence and its counterpart, that he would not be liable to be disqualified under section 35 of this Act if he were convicted of the offence, and

(b) he surrenders his licence and its counterpart to the constable or authorised person to be retained and dealt with in accordance with this Part of this Act,

the constable or authorised person must give him a fixed penalty notice in respect of the offence to which the notice under subsection (4) above relates.

(6) A notice under subsection (4) above shall give such particulars of the circumstances alleged to constitute the offence to which it relates as are necessary for giving reasonable information about the alleged offence.

(7) A licence and a counterpart of a licence surrendered in accordance with this section must be sent to the fixed penalty clerk.

KEYNOTE

The constable must be in uniform in order to give the fixed penalty notice. If the offence is endorsable (**see appendix 9, diagram 2**) the notice can only be given if the person surrenders his/her licence and the penalty incurred will not take him/her up to 12 points or beyond.

If he/she does not produce his/her licence, the person may be issued with a notice to be produced, together with the licence, within seven days at a police station.

If a person surrenders his/her licence and a notice issued under s. 54(4) at the police station within seven days, he/she must be given a fixed penalty notice so long as the total number of points on the licence will not then reach 12 or more (s. 54(5)).

If the person has not paid the fixed penalty, or given notice requesting a court hearing (under s. 55(2)), by the end of the suspended enforcement period, the police can register a sum equal to 1.5 times the amount of the penalty for enforcement against that person (s. 55(3)). Where this happens, the justices' clerk for the area where the person lives will notify the person to that effect (s. 71(6)).

If a person receives such a notification he/she can serve a statutory declaration to the effect that either:

- he/she was not the person who was given the fixed penalty notice, or
- he/she has given notice requesting a court hearing under s. 55(2).

In either case he/she must make and serve this notice within 21 days of receiving the notification from the clerk.

Under the Functions of Traffic Wardens (Amendment) Order 1986 (SI 1986 No. 1328), traffic wardens may exercise the functions conferred upon constables by Part III of the 1988 Act for certain offences under the fixed penalty system. This power is among those that can be conferred on a community support officer designated under sch. 4 to the Police Reform Act 2002 and a person accredited under sch. 5 to that Act (**see General Police Duties, chapter 2**) in relation to offences of cycling on a footway under s. 72 of the Highway Act 1835 (as to which, **see appendix 3**).

14.4.2 Where the Driver is Not Present

Section 62 of the Road Traffic Offenders Act 1988 states:

(1) Where on any occasion a constable has reason to believe in the case of any stationary vehicle that a fixed penalty offence is being or has on that occasion been committed in respect of it, he may fix a fixed penalty notice in respect of the offence to the vehicle unless the offence appears to him to involve obligatory endorsement.

KEYNOTE

Again a person may give notice during the suspended enforcement period, requesting a court hearing (s. 63(3)).

If no such notice has been given and the penalty has not been paid by the end of the suspended enforcement period, the police may serve a 'notice to owner' on the person who appears to be the owner of the vehicle (s. 63(2)).

If there is no response to the 'notice to owner', the police may register a sum equal to 1.5 times the amount of the penalty for enforcement against the person (s. 64(2)).

As in the above case where the person is present, the appropriate justices' clerk must notify the person that this has been done. The recipient can then make a statutory declaration (under sch. 4) to the effect that:

- he/she did not know of the fixed penalty notice until he/she received notification from the clerk, or
- he/she was not the owner of the vehicle at the time, or
- he/she gave notice requesting a court hearing.

If the person served with the 'notice to owner' was not in fact the owner he/she will not be liable if he/she can provide a statutory statement of ownership in relation to the vehicle (s. 64(4)).

Traffic Wardens

Under the Functions of Traffic Wardens (Amendment) Order 1986 (SI 1986/1328), traffic wardens may exercise the functions conferred upon constables by Part III of the 1988 Act for certain offences under the fixed penalty system.

14.4.3 The Offences

OFFENCE: **Making False Statements in Relation to Notice to Owner—*Road Traffic Offenders Act 1988, s. 67***
- Triable summarily • Fine

(No specific power of arrest)

The Road Traffic Offenders Act 1988, s. 67 states:

> A person who, in response to a notice to owner, provides a statement which is false in a material particular and does so recklessly or knowing it to be false in that particular is guilty of an offence.

KEYNOTE

The statement made must be false in a 'material particular', that is, in relation to some matter which is directly relevant to the information required in the notice to owner.

For the meaning of 'recklessly' and 'knowing', see Crime, chapter 1.

OFFENCE: **Removing or Interfering with Fixed Penalty Notice on Vehicle—*Road Traffic Offenders Act 1988, s. 62(2)***
- Triable summarily • Fine

(No specific power of arrest)

The Road Traffic Offenders Act 1988, s. 62 states:

> (2) A person is guilty of an offence if he removes or interferes with any notice fixed to a vehicle under this section, unless he does so by or under the authority of the driver or person in charge of the vehicle or the person liable for the fixed penalty offence in question.

KEYNOTE

Clearly if the person removing the notice can show that he/she did so on behalf of the owner or person liable for the offence, he/she would not commit this offence.

Given the procedure set out in the 1988 Act, the removal of a notice would not impede the effect of the fixed penalty system; its service can be proved by the officer under s. 79 and the notification of registration by the justices' clerk allows a defendant time to contest the allegation of the particular offence.

14.5 Conditional Offers

The introduction of automatic devices for detecting speeding and traffic signal offences (by the Road Traffic Act 1991) created the need for a method of issuing fixed penalty notices where neither of the previous circumstances applied, namely the presence of the defendant or his/her vehicle. The result was the introduction of the 'conditional offer' system.

Section 75 of the Road Traffic Offenders Act 1988 states:

(1) Where...
  (a) a constable has reason to believe that a fixed penalty offence has been committed, and
  (b) no fixed penalty notice in respect of the offence has been given under section 54 of this Act or fixed to a vehicle under section 62 of this Act, a notice under this section may be sent to the alleged offender by or on behalf of the chief officer of police.

...

(6) Where a person issues [such a notice] he must notify the justices' clerk specified in it of its issue and its terms;...

(7) A conditional offer must—
  (a) give such particulars of the circumstances alleged to constitute the offence to which it relates as are necessary for giving reasonable information about the alleged offence,
  (b) state the amount of the fixed penalty for that offence, and
  (c) state that proceedings against the alleged offender cannot be commenced in respect of that offence until the end of the period of twenty-eight days following the date on which the conditional offer was issued or such longer period as may be specified in the conditional offer.

(8) A conditional offer must indicate that if the following conditions are fulfilled, that is—
  (a) within the period of twenty-eight days following the date on which the offer was issued, or such longer period as may be specified in the offer, the alleged offender—
    (i) makes payment of the fixed penalty to the fixed penalty clerk, and
    (ii) where the offence to which the offer relates is an offence involving obligatory endorsement, at the same time delivers his licence and its counterpart to that clerk, and
  (b) where his licence and its counterpart are so delivered, that clerk is satisfied on inspecting them that, if the alleged offender were convicted of the offence, he would not be liable to be disqualified under section 35 of this Act,

  any liability to conviction of the offence shall be discharged.

KEYNOTE

A conditional offer can be sent through the post but it can only be issued where the defendant has neither been served with a fixed penalty notice in person, nor had such a notice fixed to his/her vehicle.

The offer must state that, if payment is made and—where the offence is endorsable—the defendant's licence surrendered within the specified time, any liability to conviction for the offence(s) shall be discharged.

If the defendant fails to pay the fixed penalty and, where appropriate, surrender his/her licence, the police will be notified (s. 76(5)).

If the defendant is liable to disqualification, the payment and the licence will be returned and the police notified (s. 76(4)).

---

15 Pedal cycles

15.1 Introduction

Although many provisions under road traffic legislation apply only to motor vehicles, there are a number of offences and restrictions which apply specifically to pedal cycles. Many offences have been drafted to include pedal cycles and, as mentioned in the introduction to **chapter 1**, it is important to check the exact definition of each offence or provision in every case.

For the offences of dangerous, careless and inconsiderate cycling on a road, **see chapter 2**.

This chapter sets out the more commonly-encountered legislation which applies particularly to pedal cycles.

15.2 Construction and Use

The Pedal Cycles (Construction and Use) Regulations 1983 (SI 1983 No. 1176) make provision for the construction and use of all pedal cycles including electrically assisted pedal cycles. The 1983 Regulations are treated as having been made under s. 81 of the Road Traffic Act 1988.

Breaches of the Regulations are an offence under s. 91 of the Road Traffic Offenders Act 1988.

15.2.1 Electrically Assisted Pedal Cycles

Electrically assisted pedal cycles are cycles which do not exceed 40kg kerbside weight (60kg if a tandem) and which have an electric motor that is incapable of propelling the cycle once it has reached a speed of 15 miles per hour. The best example of such a contraption is the Sinclair C5. These cycles are not motor vehicles for the purposes of the Road Traffic Act 1988 and the only qualification or authority to ride one on a road is that you are 14 years old or over (see s. 32 of the Road Traffic Act 1988).

Electrically assisted pedal cycles must, under the 1983 Regulations, be fitted with a plate showing details about the cycle and its motor; they must have a battery which does not leak, a braking system as set out in the Regulations and a device to control the motor. All of these features must be in efficient working order.

15.2.2 Pedal Cycles

Other conventional pedal cycles are subject to some of the provisions of the 1983 Regulations.

All must have a braking system which complies with reg. 7 or 8 (reg. 6). That is, they must have at least one braking system unless they are fixed-wheel cycles or cycles temporarily used in Great Britain by a visitor.

Pedal cycles with a saddle height over 635mm made on or after 1 August 1984 are required to have:

- Two independent braking systems with one acting on the front wheel(s) and one on the rear if the cycle is a free wheel cycle.
- One braking system acting on the front wheel(s) if the cycle is a fixed-wheel cycle.

Cycles made before 1 August 1984 are categorised for the purposes of their brake systems by their wheel size, with those having wheels larger than 460mm in diameter requiring similar braking systems to those above.

A constable in uniform may test a pedal cycle to ensure conformity with these requirements:

- on a road; or
- if the cycle has been involved in an accident within 48 hours before the test, on any premises provided the owner of the premises consents (reg. 11).

Selling or Supplying Unsound Pedal Cycle

It is a summary offence, under s. 81(6) of the Road Traffic Act 1988 to sell or supply (or offer to do so), a pedal cycle in such a condition that use of it on a road would contravene the 1983 Regulations.

15.3 Riding a Pedal Cycle While Unfit

OFFENCE: **Riding Cycle While Unfit—*Road Traffic Act 1988, s. 30(1)***
- Triable summarily • Fine

(No specific power of arrest)

The Road Traffic Act 1988, s. 30 states:

> (1) A person who, when riding a cycle on a road or other public place, is unfit to ride through drink or drugs (that is to say, is under the influence of drink or a drug to such an extent as to be incapable of having proper control of the cycle) is guilty of an offence.

KEYNOTE

There are no parallel provisions to those provided in respect of motorists for the requirement of a breath specimen.

There may be occasions where the person is wheeling, rather than 'riding' the cycle. In such cases the appropriate charge may be under s. 12 of the Licensing Act 1872, which creates an offence of 'being in charge of any carriage, horse or cattle on any public highway or other public place when drunk'. This offence—which carries a fine or one month's imprisonment—has been held to apply to cyclists (*Corkery* v *Carpenter* [1951] 1 KB 102). Although an ancient one, the offence is acknowledged in s. 5 of the Road Traffic Offenders Act 1988, which provides an exemption for people charged with other drink/drug related driving offences.

15.4 Cycle Racing

OFFENCE: **Cycle Racing on Public Highway—*Road Traffic Act 1988, s. 31(1)***

- Triable summarily • Fine

(No specific power of arrest)

The Road Traffic Act 1988, s. 31(1) states:

(1) A person who promotes or takes part in a race or trial of speed on a public way between cycles is guilty of an offence, unless the race or trial—
(a) is authorised, and
(b) is conducted in accordance with any conditions imposed,
by or under regulations under this section.

KEYNOTE

'Public way' means a highway (s. 31(6)); **see chapter 1.**

Section 31(4) states:

(4) Without prejudice to any other powers exercisable in that behalf, the chief officer of police may give directions with respect to the movement of, or the route to be followed by, vehicular traffic during any period, being directions which it is necessary or expedient to give in relation to that period to prevent or mitigate—

(a) congestion or obstruction of traffic, or
(b) danger to or from traffic,
in consequence of the holding of a race or trial of speed authorised by or under regulations under this section.

KEYNOTE

If injury is caused to a person by a cyclist racing or driving 'furiously' the offence under s. 35 of the Offences Against the Person Act 1861 may apply **(see chapter 2).**

There is also a specific offence of driving on footways that applies to cycles under the Highways Act 1835, s. 72.

---

16 Forgery and falsification of documents

16.1 Introduction

Much of the law regulating road traffic depends on the production and examination of documents. This chapter deals with occasions where a defendant has, or uses false documentation. When dealing with such occasions, it is important to consider the overlapping legislation which deals with deception, forgery and fraud generally (**see Crime, chapter 13**).

16.2 The Offences

OFFENCE: **Forgery of Documents—*Road Traffic Act 1988, s. 173***

- Triable either way
- Two years' imprisonment and/or a fine on indictment; fine summarily

(No specific power of arrest)

The Road Traffic Act 1988, s. 173 states:

(1) A person who, with intent to deceive—
 (a) forges, alters or uses a document or other thing to which this section applies, or
 (b) lends to, or allows to be used by, any other person a document or other thing to which this section applies, or
 (c) makes or has in his possession any document or other thing so closely resembling a document or other thing to which this section applies as to be calculated to deceive,
 is guilty of an offence.

KEYNOTE

'Forges' for this purpose means making a false document or other thing in order that it may be used as genuine (s. 173(3)).

In each of the circumstances set out in s. 173(1)(a) to (c) you must show an intention to deceive, making this a crime of 'specific intent' (**see Crime, chapter 1**).

The documents to which these offences apply are set out in s. 173(2) and include:

- licences
- test certificates
- certificates of insurance
- certificates exempting the wearing of seat belts
- international road haulage permits
- goods vehicle plates.

'Calculated to deceive' (s. 173(1)(c)) means likely to deceive. Where a defendant produces a certificate of insurance issued under a policy which has since been cancelled, this offence may be made out (see *R* v *Cleghorn* [1938] 3 All ER 398).

Although the offences under s. 173 carry no specific power of arrest, the more general offences of forgery and possessing false instruments (see Crime, chapter 13) are *arrestable offences* and can be considered in such cases. See also the power of seizure below.

OFFENCE: **False Statements and Withholding Information—*Road Traffic Act 1988, s. 174***

- Triable summarily • Fine

(No specific power of arrest)

The Road Traffic Act 1988, s. 174 states:

(1) A person who knowingly makes a false statement for the purpose—
  (a) of obtaining the grant of a licence under any Part of this Act to himself or any other person, or
  (b) of preventing the grant of any such licence, or
  (c) of procuring the imposition of a condition or limitation in relation to any such licence, or
  (d) of securing the entry or retention of the name of any person in the register of approved instructors maintained under Part V of this Act, or
  (dd) of obtaining the grant to any person of a certificate under section 133A of this Act, or
  (e) of obtaining the grant of an international road haulage permit to himself or any other person,
  is guilty of an offence.

(2) A person who, in supplying information or producing documents for the purposes either of sections 53 to 60 and 63 of this Act or of regulations made under sections 49 to 51, 61, 62 and 66(3) of this Act—
  (a) makes a statement which he knows to be false in a material particular or recklessly makes a statement which is false in a material particular, or
  (b) produces, provides, sends or otherwise makes use of a document which he knows to be false in a material particular or recklessly produces, provides, sends or otherwise makes use of a document which is false in a material particular,
  is guilty of an offence.

(3) A person who—
  (a) knowingly produces false evidence for the purposes of regulations under section 66(1) of this Act, or
  (b) knowingly makes a false statement in a declaration required to be made by the regulations,
  is guilty of an offence.

(4) A person who—
  (a) wilfully makes a false entry in any record required to be made or kept by regulations under section 74 of this Act, or
  (b) with intent to deceive, makes use of any such entry which he knows to be false,
  is guilty of an offence.

(5) A person who makes a false statement or withholds any material information for the purpose of obtaining the issue—
  (a) of a certificate of insurance or certificate of security under Part VI of this Act, or
  (b) of any document issued under regulations made by the Secretary of State in pursuance of his power under section 165(2)(a) of this Act to prescribe evidence which may be produced in lieu of a certificate of insurance or a certificate of security,
  is guilty of an offence.

KEYNOTE

There is no need to show that, in making the false statements above, the person actually gained anything or brought about the desired consequence (see e.g. *Jones* v *Meatyard* [1939] 1 All ER 140). For the interpretation of 'knowingly', 'recklessly', 'wilfully' etc., **see Crime, chapter 1.**

The offence under s. 174(4)(b) is an offence of specific intent **(see Crime, chapter 1).**

OFFENCE: **Issue of False Documents—*Road Traffic Act 1988, s. 175***

- Triable summarily • Fine

(No specific power of arrest)

The Road Traffic Act 1988, s. 175 states:

If a person issues—

(a) any such document as is referred to in section 174(5)(a) or (b) of this Act, or

(b) a test certificate or certificate of conformity (within the meaning of Part II of this Act), and the document or certificate so issued is to his knowledge false in a material particular, he is guilty of an offence.

KEYNOTE

In proving this offence you must show that the person issuing the documents knew that they were false (see *Ocean Accident etc. Co.* v *Cole* (1932) 96 JP 191).

Again it is useful to consider the offences of forgery in cases involving the issue of false documents.

Test certificates bearing a false stamp or ones which have been backdated are 'false in a material particular' (see *Murphy* v *Griffiths* [1967] 1 All ER 424).

16.2.1 Police Powers

Section 176 of the Road Traffic Act 1988 states:

(1) If a constable has reasonable cause to believe that a document produced to him—

(a) in pursuance of section 137 of this Act, or

(b) in pursuance of any of the preceding provisions of this Part of this Act,

is a document in relation to which an offence has been committed under section 173, 174 or 175 or this Act or under section 115 of the Road Traffic Regulation Act 1984, he may seize the document.

(1A) Where a licence to drive or a counterpart of any such licence or of any Community licence may be seized by a constable under subsection (1) above, he may also seize the counterpart, the licence to drive or the Community licence (as the case may be) produced with it.

(2) When a document is seized under subsection (1) above, the person from whom it was taken shall, unless—

(a) the document has been previously returned to him, or

(b) he has been previously charged with an offence under any of those sections,

be summoned before a magistrates' court... to account for his possession of the document.

(3) ...

(4) If a constable, an examiner appointed under section 66A of this Act has reasonable cause to believe that a document or plate carried on a motor vehicle or by the driver of the vehicle is a document or plate to which this subsection applies, he may seize it.

For the purposes of this subsection the power to seize includes power to detach from a vehicle.

(5) Subsection (4) above applies to a document or plate in relation to which an offence has been committed under sections 173, 174 or 175 of this Act in so far as they apply—

(a) to documents evidencing the appointment of examiners under section 66A of this Act, or

(b) to goods vehicle test certificates, plating certificates, certificates of conformity or Minister's approval certificates (within the meaning of Part II of this Act), or

(c) to plates containing plated particulars (within the meaning of that Part) or containing other particulars required to be marked on goods vehicles by sections 54 to 58 of this Act or regulations made under them, or

(d) to records required to be kept by virtue of section 74 of this Act, or

(e) to international road haulage permits.

(6) When a document or plate is seized under subsection (4) above, either the driver or owner of the vehicle shall, if the document or plate is still detained and neither of them has previously been charged with an offence in relation to the document or plate under section 173, 174 or 175 of

this Act, be summoned before a magistrates' court . . . to account for his possession of, or the presence on the vehicle of, the document or plate.

KEYNOTE

This extensive power also allows (where appropriate) for the items to be detached from the vehicle (s. 176(4)). The power under subsection (4) is restricted to officers who are authorised vehicle examiners.

The references in subsection (1) to s. 137 of the Act and s. 115 of the Road Traffic Regulation Act 1984 relate to the registration of driving instructors and the misuse of parking documents respectively.

16.3 Other Offences Involving False Records and Forgery

OFFENCE: **Forging or Altering Registration Documents—*Vehicle Excise and Registration Act 1994, ss. 44 and 45***

- Triable either way
- Two years' imprisonment and/or a fine on indictment; fine summarily

(No specific power of arrest)

The Vehicle Excise and Registration Act 1994, ss. 44 and 45 state:

44.—(1) A person is guilty of an offence if he forges, fraudulently alters, fraudulently uses, fraudulently lends or fraudulently allows to be used by another person anything to which subsection (2) applies.

(2) This subsection applies to—
(a) a vehicle licence,
(b) a trade licence,
(c) a document in the form of a licence which is issued in pursuance of regulations under this Act in respect of a vehicle which is an exempt vehicle under paragraph 19 of Schedule 2,
(d) a registration mark,
(e) a registration document, and
(f) a trade plate (including a replacement trade plate).

45.—(1) A person who in connection with—
(a) an application for a vehicle licence or a trade licence,
(b) a claim for a rebate under section 20, or
(c) an application for an allocation of registration marks,
makes a declaration which to his knowledge is either false or in any material respect misleading is guilty of an offence.

(2) A person who makes a declaration which—
(a) is required by regulations under this Act to be made in respect of a vehicle which is an exempt vehicle under paragraph 19 of Schedule 2, and
(b) to his knowledge is either false or in any material respect misleading,
is guilty of an offence.

(2A) A person who makes a declaration or statement which—
(a) is required to be made in respect of a vehicle by regulations under section 22, and
(b) to his knowledge is either false or in any material respect misleading,
is guilty of an offence.

(3) A person who—
(a) is required by [virtue of] this Act to furnish particulars relating to, or to the keeper of, a vehicle, and
(b) furnishes particulars which to his knowledge are either false or in any material respect misleading,
is guilty of an offence.

(3A) A person who, in supplying information or producing documents for the purposes of any regulations made under section 61A—
(a) makes a statement which to his knowledge is false or in any material respect misleading or recklessly makes a statement which is false or in any material respect misleading, or

(b) produces or otherwise makes use of a document which to his knowledge is false or in any material respect misleading,

is guilty of an offence.

(3B) A person who—

(a) with intent to deceive, forges, alters or uses a certificate issued by virtue of section 61A;

(b) knowing or believing that it will be used for deception lends such a certificate to another or allows another to alter or use it; or

(c) without reasonable excuse makes or has in his possession any document so closely resembling such a certificate as to be calculated to deceive,

is guilty of an offence.

KEYNOTE

'Fraudulently' means dishonestly deceiving someone who has a public duty to examine and enforce items covered by the 1988 Act (including a police officer) and there is no need to prove any intent to cause economic loss (see *R* v *Terry* [1984] AC 374).

For the law governing excise and registration of vehicles generally, see chapter 12.

OFFENCE: **Forgery of Documents Relating to Public Service Vehicles—*Public Passenger Vehicles Act 1981, s. 65(2)***

- Triable either way • Two years' imprisonment on indictment; fine summarily

(No specific power of arrest)

The Public Passenger Vehicles Act 1981, s. 65 states:

(1) This section applies to the following documents and other things, namely—

(a) a licence under Part II of this Act;

(b) a certificate of initial fitness under section 6 of this Act;

(c) a certificate under section 10 of this Act that a vehicle conforms to a type vehicle;

(d) an operator's disc under section 18 of this Act;

(e) a certificate under section 21 of this Act as to the repute, financial standing or professional competence of any person.

(2) A person who, with intent to deceive—

(a) forges or alters, or uses or lends to, or allows to be used by, any other person, a document or other thing to which this section applies, or

(b) makes or has in his possession any document or other thing so closely resembling a document or other thing to which this section applies as to be calculated to deceive, shall be liable...

KEYNOTE

'Forges' means making a false document or other thing in order that it may be used as genuine (s. 65(3)).

The expression 'calculated to deceive' means likely to do so (see para. 16.2).

For the legislation governing PSVs generally, see chapter 13.

OFFENCE: **Forgery or Alteration of Documents Relating to Goods Vehicle Operators—*Goods Vehicles (Licensing of Operators) Act 1995, s. 38***

- Triable either way
- Two years' imprisonment and/or a fine on indictment; fine summarily

(No specific power of arrest)

The Goods Vehicles (Licensing of Operators) Act 1995, s. 38 states:

(1) A person is guilty of an offence if, with intent to deceive, he—

(a) forges, alters or uses a document or other thing to which this section applies;

(b) lends to, or allows to be used by, any other person a document or other thing to which this sections applies; or

(c) makes or has in his possession any document or other thing so closely resembling a document or other thing to which this section applies as to be calculated to deceive.

(2) This section applies to the following documents and other things, namely—

(a) any operator's licence;

(b) any document, plate, mark or other thing by which, in pursuance of regulations, a vehicle is to be identified as being authorised to be used, or as being used, under an operator's licence;

(c) any document evidencing the authorisation of any person for the purposes of sections 40 and 41;

(d) any certificate of qualification under section 49; and

(e) any certificate or diploma such as is mentioned in paragraph 13(1) of Schedule 3.

KEYNOTE

'Forges' means making a false document or other thing in order that it may be used as genuine (s. 38(4)).

This offence requires proof of an intent to deceive, making it a crime of 'specific intent' (see Crime, chapter 1).

Section 41 of the 1995 Act gives a power of seizure to police officers.

For the law governing goods vehicle operators generally, see chapter 13.

OFFENCE: **Misuse of Parking Documents and Apparatus—*Road Traffic Regulation Act 1984, s. 115***

- Triable either way
- Two years' imprisonment and/or a fine on indictment; fine summarily

(No specific power of arrest)

The Road Traffic Regulation Act 1984, s. 115 states:

(1) A person shall be guilty of an offence who, with intent to deceive—

(a) uses, or lends to, or allows to be used by, any other person,—

(i) any parking device or apparatus designed to be used in connection with parking devices;

(ii) any ticket issued by a parking meter, parking device or apparatus designed to be used in connection with parking devices;

(iii) any authorisation by way of such a certificate, other means of identification or device as is referred to in any of sections 4(2), 4(3), 7(2) and 7(3) of this Act; or

(iv) any such permit or token as is referred to in section 46(2)(i) of this Act;

(b) makes or has in his possession anything so closely resembling any such thing as is mentioned in paragraph (a) above as to be calculated to deceive;...

(2) A person who knowingly makes a false statement for the purpose of procuring the grant or issue to himself or any other person of any such authorisation as is mentioned in subsection (1) above hall be guilty of an offence.

KEYNOTE

For the general offence of 'going equipped', see Crime, chapter 12.

This is an offence of 'specific intent', see Crime, chapter 1.

Appendix 1

Road Traffic Act 1988, Section 101

(1) A person is disqualified for holding or obtaining a licence to drive a motor vehicle of a class specified in the following Table if he is under the age specified in relation to it in the second column of the Table.

Table

Class of motor vehicle	Age (in years)
1. Invalid carriage	16
2. Moped	16
3. Motor bicycle	17
4. Agricultural or forestry tractor	17
5. Small vehicle	17
6. Medium-sized goods vehicle	18
7. Other motor vehicle	21

(2) The Secretary of State may by regulations provide that subsection (1) above shall have effect as if for the classes of vehicles and the ages specified in the Table in that subsection there were substituted different classes of vehicles and ages or different classes of vehicles or different ages.

(3) Subject to subsection (4) below, the regulations may—

(a) apply to persons of a class specified in or under the regulations,

(b) apply in circumstances so specified,

(c) impose conditions or create exemptions or provide for the imposition of conditions or the creation of exemptions,

(d) contain such transitional and supplemental provisions (including provisions amending section 108, 120 or 183(5) of this Act) as the Secretary of State considers necessary or expedient.

(4) For the purpose of defining the class of persons to whom, the class of vehicles to which, the circumstances in which or the conditions subject to which regulations made by virtue of subsection (2) above are to apply where an approved proved training scheme for drivers is in force, it is sufficient for the regulations to refer to a document which embodies the terms (or any of the terms) of the scheme or to a document which is in force in pursuance of the scheme.

(5) In subsection (4) above—

'approved' means approved for the time being by the Secretary of State for the purpose of the regulations,

'training scheme for drivers' means a scheme for training persons to drive vehicles of a class in relation to which the age which is in force under this section (but apart from any such scheme) is 21 years, but no approved training scheme for drivers shall be amended without the approval of the Secretary of State.

Appendix 2

Road Traffic Regulation Act 1984, Schedule 6

Schedule 6 Speed Limits for Vehicles of Certain Classes

PART I VEHICLES FITTED WITH PNEUMATIC TYRES ON ALL WHEELS
(see application provisions below the following Table)

TABLE

1	2			
Item No.	Class of Vehicle	Maximum speed (in miles per hour) while vehicle is being driven on:		
		(a) Motorway	(b) Dual carriageway road not being a motorway	(c) Other road
1.	A passenger vehicle, motor caravan or dual-purposevehicle not drawing a trailer being a vehicle with an unladen weight exceeding 3.05 tonnes or adapted to carry more than 8 passengers:			
	(i) if not exceeding 12 metres in overall length	70	60	50
	(ii) if exceeding 12 metres in overall length	60	60	50
2.	An invalid carriage	not applicable	20	20
3.	A passenger vehicle, motor caravan, car-derived van or dual-purpose vehicle drawing one trailer	60	60	50
4.	A passenger vehicle, motor caravan, car-derived van or dual-purpose vehicle drawing more than one trailer	40	20	20
5.	(1) A goods vehicle having a maximum laden weight not exceeding 7.5 tonnes and which is not— (a) an articulated vehicle, or (b) drawing a trailer, or (c) a car-derived van	70	60	50
	(2) A goods vehicle which is— (a) (i) an articulated vehicle having a maximum laden weight not exceeding 7.5 tonnes, or (ii) a motor vehicle, other			

Table (*continued*)

1	2			
Item No.	Class of Vehicle	Maximum speed (in miles per hour) while vehicle is being driven on:		
		(a) Motorway	(b) Dual carriageway road not being a motorway	(c) Other road
	than a car-derived van, which is drawing one trailer where the aggregate maximum laden weight of the motor vehicle and the trailer does not exceed 7.5 tonnes	60	60	50
	(b) (i) an articulated vehicle having a maximum laden weight exceeding 7.5 tonnes, (ii) a motor vehicle having a maximum laden weight exceeding 7.5 tonnes and not drawing a trailer, or (iii) a motor vehicle drawing one trailer where the aggregate maximum laden weight of the motor vehicle and the trailer exceeds 7.5 tonnes	60	50	40
	(c) a motor vehicle, other than a car-derived van, drawing more than one trailer	40	20	20
6.	A motor tractor (other than an industrial tractor), a light locomotive or a heavy locomotive—			
	(a) if the provisions about springs and wings as specified in paragraph 3 of Part IV of this Schedule are complied with, and the vehicle is not drawing a trailer, or if those provisions are complied with and the vehicle is drawing one trailer, which also complies with those provisions	40	30	30
	(b) in any other case	20	20	20
7.	A works truck	18	18	18
8.	An industrial tractor	not applicable	18	18
9.	An agricultural motor vehicle	40	40	40

Application

This Part applies only to motor vehicles, not being track-laying vehicles, every wheel of which is fitted with a pneumatic tyre and to such vehicles drawing one or more trailers, not being track-laying vehicles, every wheel of which is fitted with a pneumatic tyre.

Appendix 3

Schedules 2 and 3 to the Road Traffic Offenders Act 1988

Schedule 2 to the Road Traffic Offenders Act 1988

SCHEDULE 2
PROSECUTION AND PUNISHMENT OF OFFENCES

PART I
OFFENCES UNDER THE TRAFFIC ACTS

(1) Provision creating offence	(2) General nature of offence	(3) Mode of prosecution	(4) Punishment	(5) Disqualification	(6) Endorsement	(7) Penalty points
		Offences under the Road Traffic Regulation Act 1984				
RTRA section 5	Contravention of traffic regulation order.	Summarily.	Level 3 on the standard scale.			
RTRA section 8	Contravention of order regulating traffic in Greater London.	Summarily.	Level 3 on the standard scale.			
RTRA section 11	Contravention of experimental traffic order.	Summarily.	Level 3 on the standard scale.			
RTRA section 13	Contravention of experimental traffic scheme in Greater London.	Summarily.	Level 3 on the standard scale.			
RTRA section 16(1)	Contravention of temporary prohibition or restriction.	Summarily.	Level 3 on the standard scale.	Discretionary if committed in respect of a speed restriction.	Obligatory if committed in respect of a speed restriction.	3–6 or 3 (fixed penalty).
RTRA section 16C(1)	Contravention of prohibition or restriction relating to relevant event.	Summarily.	Level 3 on the standard scale.			
RTRA section 17(4)	Use of special road contrary to scheme or regulations.	Summarily.	Level 4 on the standard scale.	Discretionary if committed in respect of a motor vehicle otherwise than by unlawfully stopping or allowing the vehicle to remain at rest on a part of	Obligatory if committed as mentioned in the entry in column 5.	3–6 or 3 (fixed penalty) if committed in respect of a speed restriction, 3 in any other case.

(1) Provision creating offence	(2) General nature of offence	(3) Mode of prosecution	(4) Punishment	(5) Disqualification	(6) Endorsement	(7) Penalty points
Offences under the Road Traffic Regulation Act 1984—continued						
				a special road on which vehicles are in certain circumstances permitted to remain at rest.		
RTRA section 18(3)	One-way traffic on trunk road.	Summarily.	Level 3 on the standard scale.			
RTRA section 20(5)	Contravention of prohibition or restriction for roads of certain classes.	Summarily.	Level 3 on the standard scale.			
RTRA section 25(5)	Contravention of pedestrian crossing regulations.	Summarily.	Level 3 on the standard scale.	Discretionary if committed in respect of a motor vehicle.	Obligatory if committed in respect of a motor vehicle.	3
RTRA section 28(3)	Not stopping at school crossing.	Summarily.	Level 3 on the standard scale.	Discretionary if committed in respect of a motor vehicle.	Obligatory if committed in respect of a motor vehicle.	3
RTRA section 29(3)	Contravention of order relating to street playground.	Summarily.	Level 3 on the standard scale.	Discretionary if committed in respect of a motor vehicle.	Obligatory if committed in respect of a motor vehicle.	2
RTRA section 35A(1)	Contravention of order as to use of parking place.	Summarily.	(a) Level 3 on the standard scale in the case of an offence committed by a person in a street parking place reserved for disabled persons' vehicles or in an off-street parking place reserved for such vehicles, where that person would not have been guilty of that offence if the motor vehicle in respect of which it was committed had been a disabled person's vehicle.			

(1) Provision creating offence	(2) General nature of offence	(3) Mode of prosecution	(4) Punishment	(5) Disqualification	(6) Endorsement	(7) Penalty points
Offences under the Road Traffic Regulation Act 1984—continued						
			(b) Level 2 on the standard scale in any other case.			
RTRA section 35A(2)	Misuse of apparatus for collecting charges or of parking device or connected apparatus.	Summarily.	Level 3 on the standard scale.			
RTRA section 35A(5)	Plying for hire in parking place.	Summarily.	Level 2 on the standard scale.			
RTRA section 43(5)	Unauthorised disclosure of information in respect of licensed parking place.	Summarily.	Level 3 on the standard scale.			
RTRA section 43(10)	Failure to comply with term or conditions of licence to operate parking place.	Summarily.	Level 3 on the standard scale.			
RTRA section 43(12)	Operation of public off-street parking place without licence.	Summarily.	Level 5 on the standard scale.			
RTRA section 47(1)	Contraventions relating to designated parking places.	Summarily.	(a) Level 3 on the standard scale in the case of an offence committed by a person in a street parking place reserved for disabled persons' vehicles where that person would not have been guilty of the offence if the motor vehicle in respect of which it was committed had been a disabled person's vehicle. (b) Level 2 in any other case.			
RTRA section 47(3)	Tampering with parking meter.	Summarily.	Level 3 on the standard scale.			
RTRA section 52(1)	Misuse of parking device.	Summarily.	Level 2 on the standard scale.			
RTRA section 53(5)	Contravention of certain provisions of designation orders.	Summarily.	Level 3 on the standard scale.			
RTRA section 53(6)	Other contraventions of designation orders.	Summarily.	Level 2 on the standard scale.			

(1) Provision creating offence	(2) General nature of offence	(3) Mode of prosecution	(4) Punishment	(5) Disqualification	(6) Endorsement	(7) Penalty points
	Offences under the Road Traffic Regulation Act 1984—continued					
RTRA section 61(5)	Unauthorised use of loading area.	Summarily.	Level 3 on the standard scale.			
RTRA section 88(7)	Contravention of minimum speed limit.	Summarily.	Level 3 on the standard scale.			
RTRA section 89(1)	Exceeding speed limit.	Summarily.	Level 3 on the standard scale.	Discretionary.	Obligatory.	3–6 or 3 (fixed penalty).
RTRA section 104(5)	Interference with notice as to immobilisation device.	Summarily.	Level 2 on the standard scale.			
RTRA section 104(6)	Interference with immobilisation device.	Summarily.	Level 3 on the standard scale.			
RTRA section 105(5)	Misuse of disabled person's badge (immobilisation devices).	Summarily.	Level 3 on the standard scale.			
RTRA section 108(2) (or that subsection as modified by section 109(2) and (3)).	Non-compliance with notice (excess charge).	Summarily.	Level 3 on the standard scale.			
RTRA section 108(3) (or that subsection as modified by section 109(2) and (3)).	False response to notice (excess charge).	Summarily.	Level 5 on the standard scale.			
RTRA section 112(4)	Failure to give information as to identity of driver.	Summarily.	Level 3 on the standard scale.			
RTRA section 115(1)	Mishandling or faking parking documents.	(a) Summarily. (b) On indictment.	(a) The statutory maximum. (b) 2 years.			
RTRA section 115(2)	False statement for procuring authorisation.	Summarily.	Level 4 on the standard scale.			
RTRA section 116(1)	Non-delivery of suspect document or article.	Summarily.	Level 3 on the standard scale.			
RTRA section 117	Wrongful use of disabled person's badge.	Summarily.	Level 3 on the standard scale.			
RTRA section 129(3)	Failure to give evidence at inquiry.	Summarily.	Level 3 on the standard scale.			
	Offences under the Road Traffic Act 1988					
RTA section 1	Causing death by dangerous driving.	On indictment.	10 years.	Obligatory.	Obligatory.	3–11
RTA section 2	Dangerous Driving.	(a) Summarily. (b) On indictment.	(a) 6 months or the statutory maximum or both. (b) 2 years or a fine or both.	Obligatory.	Obligatory.	3–11
RTA section 3	Careless, and inconsiderate, driving.	Summarily.	Level 4 on the standard scale.	Discretionary.	Obligatory.	3–9
RTA section 3A	Causing death by careless driving	On indictment.	10 years or a fine or both.	Obligatory.	Obligatory.	3–11

(1) Provision creating offence	(2) General nature of offence	(3) Mode of prosecution	(4) Punishment	(5) Disqualification	(6) Endorsement	(7) Penalty points
			Offences under the Road Traffic Act 1988—continued			
	when under influence of drink or drugs.					
RTA section 4(1)	Driving or attempting to drive when unfit to drive through drink or drugs.	Summarily.	6 months or level 5 on the standard scale or both.	Obligatory.	Obligatory.	3–11
RTA section 4(2)	Being in charge of a mechanically propelled vehicle when unfit to drive through drink or drugs.	Summarily.	3 months or level 4 on the standard scale or both.	Discretionary.	Obligatory.	10
RTA section 5(1)(a)	Driving or attempting to drive with excess alcohol in breath, blood or urine.	Summarily.	6 months or level 5 on the standard scale or both.	Obligatory.	Obligatory.	3–11
RTA section 5(1)(b)	Being in charge of a motor vehicle with excess alcohol in breath, blood or urine.	Summarily.	3 months or level 4 on the standard scale or both.	Discretionary.	Obligatory.	10
RTA section 6	Failing to provide a specimen of breath for a breath test.	Summarily.	Level 3 on the standard scale.	Discretionary.	Obligatory.	4
RTA section 7	Failing to provide specimen for analysis or laboratory test.	Summarily.	(a) Where the specimen was required to ascertain ability to drive or proportion of alcohol at the time offender was driving or attempting to drive, 6 months or level 5 on the standard scale or both.	(a) Obligatory in case mentioned in column 4(a).	Obligatory.	(a) 3–11 in case mentioned in column 4(a).
			(b) In any other case, 3 months or level 4 on the standard scale or both.	(b) Discretionary in any other case.		(b) 10 in any other case.
RTA section 7A	Failing to allow specimen to be subjected to laboratory test.	Summarily.	(a) Where the test would be for ascertaining ability to drive or proportion of alcohol at the time offender was driving	(a) Obligatory in the case mentioned in column 4(a).	Obligatory	3–11, in case mentioned in column 4(a)

(1) Provision creating offence	(2) General nature of offence	(3) Mode of prosecution	(4) Punishment	(5) Disqualification	(6) Endorsement	(7) Penalty points
		Offences under the Road Traffic Act 1988—continued				
			or attempting to drive, 6 months or level 5 on the standard scale or both. (b) In any other case, 3 months or level 4 on the standard scale or both.	(b) Discretionary in any other case.	10, in any other case.	
RTA section 12	Motor racing and speed trials on public ways.	Summarily.	Level 4 on the standard scale.	Obligatory.	Obligatory.	3–11
RTA section 13	Other unauthorised or irregular competitions or trials on public ways.	Summarily.	Level 3 on the standard scale.			
RTA section 14	Driving or riding in a motor vehicle in contravention of regulations requiring wearing of seat belts.	Summarily.	Level 2 on the standard scale.			
RTA section 15(2)	Driving motor vehicle with child not wearing seat belt.	Summarily.	Level 2 on the standard scale.			
RTA section 15(4)	Driving motor vehicle with child in rear not wearing a seat belt.	Summarily.	Level 1 on the standard scale.			
RTA section 15A(3) or (4)	Selling etc. in certain circumstances equipment as conducive to the safety of children in motor vehicles.	Summarily.	Level 3 on the standard scale.			
RTA section 16	Driving or riding motor cycles in contravention of regulations requiring wearing of protective headgear.	Summarily.	Level 2 on the standard scale.			
RTA section 17	Selling, etc., helmet not of the prescribed type as helmet for affording protection for motor cyclists.	Summarily.	Level 3 on the standard scale.			
RTA section 18(3)	Contravention of regulations with respect to use of head-worn appliances on motor cycles.	Summarily.	Level 2 on the standard scale.			
RTA section 18(4)	Selling, etc., appliance not of prescribed type	Summarily.	Level 3 on the standard scale.			

(1) Provision creating offence	(2) General nature of offence	(3) Mode of prosecution	(4) Punishment	(5) Disqualification	(6) Endorsement	(7) Penalty points
	Offences under the Road Traffic Act 1988—continued					
	as approved for use on motor cycles.					
RTA section 19	Prohibition of parking of heavy commercial vehicles on verges, etc.	Summarily.	Level 3 on the standard scale.			
RTA section 21	Driving or parking on cycle track.	Summarily.	Level 3 on the standard scale.			
RTA section 22	Leaving vehicles in dangerous positions.	Summarily.	Level 3 on the standard scale.	Discretionary if committed in respect of a motor vehicle.	Obligatory if committed in respect of a motor vehicle.	3
RTA section 22A	Causing danger to road users.	(a) Summarily. (b) On indictment.	(a) 6 months or the statutory maximum or both. (b) 7 years or a fine or both.			
RTA section 23	Carrying passenger on motor-cycle contrary to section 23.	Summarily.	Level 3 on the standard scale.	Discretionary.	Obligatory.	3
RTA section 24	Carrying passenger on bicycle contrary to section 24.	Summarily.	Level 1 on the standard scale.			
RTA section 25	Tampering with motor vehicles.	Summarily.	Level 3 on the standard scale.			
RTA section 26	Holding or getting on to vehicle, etc., in order to be towed or carried.	Summarily.	Level 1 on the standard scale.			
RTA section 27	Dogs on designated roads without being held on lead.	Summarily.	Level 1 on the standard scale.			
RTA section 28	Dangerous cycling.	Summarily.	Level 4 on the standard scale.			
RTA section 29	Careless, and inconsiderate, cycling.	Summarily.	Level 3 on the standard scale.			
RTA section 30	Cycling when unfit through drink or drugs.	Summarily.	Level 3 on the standard scale.			
RTA section 31	Unauthorised or irregular cycle racing or trials of speed on public ways.	Summarily.	Level 1 on the standard scale.			
RTA section 32	Contravening prohibition on persons under 14 driving electrically assisted pedal cycles.	Summarily.	Level 2 on the standard scale.			
RTA section 33	Unauthorised motor vehicle trial on footpaths or bridleways.	Summarily.	Level 3 on the standard scale.			
RTA section 34	Driving mechanically propelled vehicles elsewhere than on roads.	Summarily.	Level 3 on the standard scale.			

(1) Provision creating offence	(2) General nature of offence	(3) Mode of prosecution	(4) Punishment	(5) Disqualification	(6) Endorsement	(7) Penalty points
		Offences under the Road Traffic Act 1988—continued				
RTA section 35	Failing to comply with traffic directions.	Summarily.	Level 3 on the standard scale.	Discretionary, if committed in respect of a motor vehicle by failure to comply with a direction of a constable or traffic warden.	Obligatory if committed as described in column 5.	3
RTA section 36	Failing to comply with traffic signs.	Summarily.	Level 3 on the standard scale.	Discretionary, if committed in respect of a motor vehicle by failure to comply with an indication given by a sign specified for the purposes of this paragraph in regulations under RTA section 36.	Obligatory if committted as described in column 5.	3
RTA section 37	Pedestrian failing to stop when directed by constable regulating traffic.	Summarily.	Level 3 on the standard scale.			
RTA section 40A	Using vehicle in dangerous condition etc.	Summarily.	(a) Level 5 on the standard scale if committed in respect of a goods vehicle or a vehicle adapted to carry more than eight passengers. (b) Level 4 on the standard scale in any other case.	Discretionary.	Obligatory.	3
RTA section 41A	Breach of requirement as to brakes, steering-gear or tyres.	Summarily.	(a) Level 5 on the standard scale if committed in respect of a goods vehicle or a vehicle adapted to carry more than eight passengers. (b) Level 4 on the standard scale in any other case.	Discretionary.	Obligatory.	3

(1) Provision creating offence	(2) General nature of offence	(3) Mode of prosecution	(4) Punishment	(5) Disqualification	(6) Endorsement	(7) Penalty points
		Offences under the Road Traffic Act 1988—continued				
RTA section 41B	Breach of requirement as to weight: goods and passenger vehicles.	Summarily.	Level 5 on the standard scale.			
RTA section 42	Breach of other construction and use requirements.	Summarily.	(a) Level 4 on the standard scale if committed in respect of a goods vehicle or a vehicle adapted to carry more than eight passengers. (b) Level 3 on the standard scale in any other case.			
RTA section 47	Using, etc., vehicle without required test certificate being in force.	Summarily.	(a) Level 4 on the standard scale in the case of a vehicle adapted to carry more than eight passengers. (b) Level 3 on the standard scale in any other case.			
Regulations under RTA section 49 made by virtue of section 51(2)	Contravention of requirement of regulations (which is declared by regulations to be an offence) that driver of goods vehicle being tested be present throughout test or drive, etc., vehicle as and when directed.	Summarily.	Level 3 on the standard scale.			
RTA section 53(1)	Using, etc., goods vehicle without required plating certificate being in force.	Summarily.	Level 3 on the standard scale.			
RTA section 53(2)	Using, etc., goods vehicle without required goods vehicle test certificate being in force.	Summarily.	Level 4 on the standard scale.			

(1) Provision creating offence	(2) General nature of offence	(3) Mode of prosecution	(4) Punishment	(5) Disqualification	(6) Endorsement	(7) Penalty points
		Offences under the Road Traffic Act 1988—continued				
RTA section 53(3)	Using, etc., goods vehicle where Secretary of State is required by regulations under section 49 to be notified of an alteration to the vehicle or its equipment but has not been notified.	Summarily.	Level 3 on the standard scale.			
Regulations under RTA section 61 made by virtue of subsection (4)	Contravention of requirement of regulations (which is declared by regulations to be an offence) that driver of goods vehicle being tested after notifiable alteration be present throughout test and drive, etc., vehicle as and when directed.	Summarily.	Level 3 on the standard scale.			
RTA section 63(1)	Using, etc., goods vehicle without required certificate being in force showing that it complies with type approval requirements applicable to it.	Summarily.	Level 4 on the standard scale.			
RTA section 63(2)	Using, etc., certain goods vehicles for drawing trailer when plating certificate does not specify maximum laden weight for vehicle and trailer.	Summarily.	Level 3 on the standard scale.			
RTA section 63(3)	Using, etc., goods vehicle where Secretary of State is required to be notified under section 59 of alteration to it or its equipment but has not been notified.	Summarily.	Level 3 on the standard scale.			
RTA section 64	Using goods vehicle with unauthorised weights as well as authorised weights marked on it.	Summarily.	Level 3 on the standard scale.			

(1) Provision creating offence	(2) General nature of offence	(3) Mode of prosecution	(4) Punishment	(5) Disqualification	(6) Endorsement	(7) Penalty points
	Offences under the Road Traffic Act 1988—continued					
RTA section 64A	Failure to hold EC certificate of conformity for unregistered light passenger vehicle or motor cycle.	Summarily.	Level 3 on the standard scale.			
RTA section 65	Supplying vehicle or vehicle part without required certificate being in force showing that it complies with type approval requirements applicable to it.	Summarily.	Level 5 on the standard scale.			
RTA section 65A	Light passenger vehicles and motor cycles not to be sold without EC certificate of conformity.	Summarily.	Level 5 on the standard scale.			
RTA section 67	Obstructing testing of vehicle by examiner on road or failing to comply with requirements of RTA section 67 or schedule 2.	Summarily.	Level 3 on the standard scale.			
RTA section 68	Obstructing inspection, etc., of vehicle by examiner or failing to comply with requirement to take vehicle for inspection.	Summarily.	Level 3 on the standard scale.			
RTA section 71	Driving, etc., vehicle in contravention of prohibition on driving it as being unfit for service, or refusing, neglecting or otherwise failing to comply with direction to remove a vehicle found overloaded.	Summarily.	Level 5 on the standard scale.			
RTA section 74	Contravention of regulations requiring goods vehicle operator to inspect, and keep records of inspection of, goods vehicles.	Summarily.	Level 3 on the standard scale.			
RTA section 75	Selling, etc., unroadworthy vehicle or trailer or altering vehicle or trailer so as to make it unroadworthy.	Summarily.	Level 5 on the standard scale.			

(1) Provision creating offence	(2) General nature of offence	(3) Mode of prosecution	(4) Punishment	(5) Disqualification	(6) Endorsement	(7) Penalty points
		Offences under the Road Traffic Act 1988—continued				
RTA section 76(1)	Fitting of defective or unsuitable vehicle parts.	Summarily.	Level 5 on the standard scale.			
RTA section 76(3)	Supplying defective or unsuitable vehicle parts.	Summarily.	Level 4 on the standard scale.			
RTA section 76(8)	Obstructing examiner testing vehicles to ascertain whether defective or unsuitable part has been fitted, etc.	Summarily.	Level 3 on the standard scale.			
RTA section 77	Obstructing examiner testing condition of used vehicle at sale rooms, etc.	Summarily.	Level 3 on the standard scale.			
RTA section 78	Failing to comply with requirement about weighing motor vehicle or obstructing authorised person.	Summarily.	Level 5 on the standard scale.			
RTA section 81	Selling, etc., pedal cycle in contravention of regulations as to brakes, bells, etc.	Summarily.	Level 3 on the standard scale.			
RTA section 83	Selling, etc., wrongly made tail lamps or reflectors.	Summarily.	Level 5 on the standard scale.			
RTA section 87(1)	Driving otherwise than in accordance with a licence.	Summarily.	Level 3 on the standard scale.	Discretionary in a case where the offender's driving would not have been in accordance with any licence that could have been granted to him.	Obligatory in the case mentioned in column 5.	3–6
RTA section 87(2)	Causing or permitting a person to drive otherwise than in accordance with a licence.	Summarily.	Level 3 on the standard scale.			
RTA section 92(7C)	Failure to deliver licence revoked by virtue of section 92(7A) and counterpart to Secretary of State.	Summarily.	Level 3 on the standard scale.			
RTA section 92(10)	Driving after making false declaration as to physical fitness.	Summarily.	Level 4 on the standard scale.	Discretionary.	Obligatory.	3–6
RTA section 93(3)	Failure to deliver revoked licence and counterpart to Secretary of State.	Summarily.	Level 3 on the standard scale.			

(1) Provision creating offence	(2) General nature of offence	(3) Mode of prosecution	(4) Punishment	(5) Disqualification	(6) Endorsement	(7) Penalty points
		Offences under the Road Traffic Act 1988—continued				
RTA section 94(3) and that subsection as applied by RTA section 99D	Failure to notify Secretary of State of onset of, or deterioration in, relevant or prospective disability.	Summarily.	Level 3 on the standard scale.			
RTA section 94(3A) and that subsection as applied by RTA section 99D(b)	Driving after such a failure.	Summarily.	Level 3 on the standard scale.	Discretionary.	Obligatory.	3–6
RTA section 94A	Driving after refusal of licence under section 92(3), revocation under section 93 or service of a notice under section 99C.	Summarily.	6 months or level 5 on the standard scale or both.	Discretionary.	Obligatory.	3–6
RTA section 96	Driving with uncorrected defective eyesight, or refusing to submit to test of eyesight.	Summarily.	Level 3 on the standard scale.	Discretionary.	Obligatory.	3
RTA section 99(5)	Driving licence holder failing, to surrender licence and counterpart.	Summarily.	Level 3 on the standard scale.			
RTA section 99B(11)	Driving after failure to comply with a requirement under section 99B(6), (7) or (10).	Summarily.	Level 3 on the standard scale.			
RTA section 99C(4)	Failure to deliver Community licence to Secretary of State when required by notice under section 99C.	Summarily.	Level 3 on the standard scale.			
RTA section 103(1)(a)	Obtaining driving licence while disqualified.	Summarily.	Level 3 on the standard scale.			
RTA section 103(1)(b)	Driving while disqualified.	(a) Summarily, in England and Wales. (b) Summarily, in Scotland. (c) On indictment, in Scotland.	(a) 6 months or level 5 on the standard scale or both. (b) 6 months or the statutory maximum or both. (c) 12 months or a fine or both.	Discretionary.	Obligatory.	6

(1) Provision creating offence	(2) General nature of offence	(3) Mode of prosecution	(4) Punishment	(5) Disqualification	(6) Endorsement	(7) Penalty points
		Offences under the Road Traffic Act 1988—continued				
RTA section 109	Failing to produce to court Northern Ireland driving licence and its counterpart.	Summarily.	Level 3 on the standard scale.			
RTA section 114	Failing to comply with conditions of LGV, PCV licence or LGV Community licence, or causing or permitting person under 21 to drive LGV or PCV in contravention of such conditions.	Summarily.	Level 3 on the standard scale.			
RTA section 115A(4)	Failure to deliver LGV or PCV Community licence when required by notice under section 115A.	Summarily.	Level 3 on the standard scale.			
RTA section 118	Failing to surrender revoked or suspended LGV or PCV licence and counterpart.	Summarily.	Level 3 on the standard scale.			
Regulations made by virtue of RTA section 120(5)	Contravention of provision of regulations (which is declared by regulations to be an offence) about LGV or PCV drivers' licences or LGV or PCV Community licence.	Summarily.	Level 3 on the standard scale.			
RTA section 123(4)	Giving of paid driving instruction by unregistered and unlicensed persons or their employers.	Summarily.	Level 4 on the standard scale.			
RTA section 123(6)	Giving of paid instruction without there being exhibited on the motor car a certificate of registration or a licence under RTA part V.	Summarily.	Level 3 on the standard scale.			
RTA section 125A(4)	Failure, on application for registration as disabled driving instructor, to notify Registrar of onset of, or deterioration in, relevant or prospective disability.	Summarily.	Level 3 on the standard scale.			

(1) Provision creating offence	(2) General nature of offence	(3) Mode of prosecution	(4) Punishment	(5) Disqualification	(6) Endorsement	(7) Penalty points
	Offences under the Road Traffic Act 1988—continued					
RTA section 133C(4)	Failure by registered or licensed disabled driving instructor to notify Registrar of onset of, or deterioration in, relevant or prospective disability.	Summarily.	Level 3 on the standard scale.			
RTA section 133D	Giving of paid driving instruction by disabled persons or their employers without emergency control certificate or in unauthorised motor car.	Summarily.	Level 3 on the standard scale.			
RTA section 135	Unregistered instructor using title or displaying badge, etc., prescribed for registered instructor, or employer using such title, etc., in relation to his unregistered instructor or issuing misleading advertisement, etc.	Summarily.	Level 4 on the standard scale.			
RTA section 136	Failure of instructor to surrender to Registrar certificate or licence.	Summarily.	Level 3 on the standard scale.			
RTA section 137	Failing to produce certificate of registration or licence as driving instructor.	Summarily.	Level 3 on the standard scale.			
RTA section 143	Using motor vehicle while uninsured or unsecured against third-party risks.	Summarily.	Level 5 on the standard scale.	Discretionary.	Obligatory.	6–8
RTA section 147	Failing to surrender certificate of insurance or security to insurer on cancellation or to make statutory declaration of loss or destruction.	Summarily.	Level 3 on the standard scale.			
RTA section 154	Failing to give information, or wilfully making a false statement, as to insurance or security when claim made.	Summarily.	Level 4 on the standard scale.			
RTA section 163	Failing to stop motor vehicle or cycle when required by constable.	Summarily.	Level 3 on the standard scale.			

(1) Provision creating offence	(2) General nature of offence	(3) Mode of prosecution	(4) Punishment	(5) Disqualification	(6) Endorsement	(7) Penalty points
		Offences under the Road Traffic Act 1988—continued				
RTA section 164	Failing to produce driving licence and its counterpart or to state date of birth, or failing to provide the Secretary of State with evidence of date of birth, etc.	Summarily.	Level 3 on the standard scale.			
RTA section 165	Failing to give certain names and addresses or to produce certain documents.	Summarily.	Level 3 on the standard scale.			
RTA section 168	Refusing to give, or giving false, name and address in case of reckless, careless or inconsiderate driving or cycling.	Summarily.	Level 3 on the standard scale.			
RTA section 169	Pedestrian failing to give constable his name and address after failing to stop when directed by constable controlling traffic.	Summarily.	Level 1 on the standard scale.			
RTA section 170(4)	Failing to stop after accident and give particulars or report accident.	Summarily.	Six months or level 5 on the standard scale or both.	Discretionary.	Obligatory.	5–10
RTA section 170(7)	Failure by driver, in case of accident involving injury to another, to produce evidence of insurance or security or to report accident.	Summarily.	Level 3 on the standard scale.			
RTA section 171	Failure by owner of motor vehicle to give police information for verifying compliance with requirement of compulsory insurance or security.	Summarily.	Level 4 on the standard scale.			
RTA section 172	Failure of person keeping vehicle and others to give police information as to identity of driver, etc., in the case of certain offences.	Summarily.	Level 3 on the standard scale.	Discretionary if committed otherwise than than by virtue of subsection (5) or (11).	Obligatory if committed otherwise than than by virtue of subsection (5) or (11).	3
RTA section 173	Forgery, etc., of licences, counterparts of Community licences, certificates of insurance and other documents and things.	(a) Summarily.	(a) The statutory maximum.			
		(b) On indictment.	(b) 2 years.			

(1) Provision creating offence	(2) General nature of offence	(3) Mode of prosecution	(4) Punishment	(5) Disqualification	(6) Endorsement	(7) Penalty points
		Offences under the Road Traffic Act 1988—continued				
RTA section 174	Making certain false statements, etc., and withholding certain material information.	Summarily.	Level 4 on the standard scale.			
RTA section 175(1)	Issuing false documents.	Summarily.	Level 4 on the standard scale.			
RTA section 175(2)	Falsely amending certificate of conformity.	Summarily.	Level 4 on the standard scale.			
RTA section 177	Impersonation of, or of person employed by, authorised examiner.	Summarily.	Level 3 on the standard scale.			
RTA section 178	[Scotland.]					
RTA section 180	Failing to attend, give evidence or produce documents to, inquiry held by Secretary of State, etc.	Summarily.	Level 3 on the standard scale.			
RTA section 181	Obstructing inspection of vehicles after accident.	Summarily.	Level 3 on the standard scale.			
RTA schedule 1 paragraph 6	Applying warranty to equipment, protective helmet, appliance or information in defending proceedings under RTA section 15A, 17 or 18(4) where no warranty given, or applying false warranty.	Summarily.	Level 3 on the standard scale.			
Section 25 of this Act.	Failing to give information as to date of birth or sex to court or to provide Secretary of State with evidence of date of birth, etc.	Summarily.	Level 3 on the standard scale.			
Section 26 of this Act.	Failing to produce driving licence and its counterpart to court making order for interim disqualification.	Summarily.	Level 3 on the standard scale.			
Section 27 of this Act.	Failing to produce licence and its counterpart to court for endorsement on conviction of offence involving obligatory endorsement or on committal for sentence, etc., for offence involving	Summarily.	Level 3 on the standard scale.			

(1) Provision creating offence	(2) General nature of offence	(3) Mode of prosecution	(4) Punishment	(5) Disqualification	(6) Endorsement	(7) Penalty points
Section 45 of this Act.	Applying for or obtaining licence without giving particulars of current endorsement	Summarily.	Level 3 on the standard scale.			
	Offences under the Road Traffic Act 1988—continued					
	obligatory or discretionary disqualification when no interim disqualification ordered.					
Section 62 of this Act.	Removing fixed penalty notice fixed to vehicle.	Summarily.	Level 2 on the standard scale.			
Section 67 of this Act.	False statement in response to notice to owner.	Summarily.	Level 5 on the standard scale.			
	Offences under the Road Traffic (Driver Licensing and Information Systems) Act 1989					
RTA 1989 section 1(5)	Failure of holder of existing HGV or PSV driver's licence to surrender it upon revocation or surrender of his existing licence under part III of RTA.	Summarily.	Level 3 on the standard scale.			
RTA 1989 schedule 1, para. 3	Failing to comply with conditions of existing HGV driver's licence, or causing or permitting a person under 21 to drive HGV in contravention of such conditions.	Summarily.	Level 3 on the standard scale.			
RTA 1989 schedule 1, para. 8(2)	Contravention of provision of regulations (which is declared by regulations to be an offence) about existing HGV or PSV drivers' licences.	Summarily.	Level 3 on the standard scale.			
RTA 1989 schedule 1, para. 10(4)	Taking PSV test before applying for licence or within prescribed period afterwards.	Summarily.	Level 3 on the standard scale.			
RTA 1989 schedule 1, para. 10(5)	Taking PSV test after refusal of a licence.	Summarily.	Level 3 on the standard scale.			

PART II
OTHER OFFENCES

(1) Offence	(2) Disqualification	(3) Endorsement	(4) Penalty points
Manslaughter or, in Scotland, culpable homicide by the driver of a motor vehicle.	Obligatory.	Obligatory.	3–11
An offence under section 12A of the Theft Act 1968 (aggravated vehicle-taking).	Obligatory.	Obligatory.	3–11
Stealing or attempting to steal a motor vehicle.	Discretionary.		
An offence or attempt to commit an offence in respect of a motor vehicle under section 12 of the Theft Act 1968 (taking conveyance without consent of owner etc. or, knowing it has been so taken, driving it or allowing oneself to be carried in it).	Discretionary.		
An offence under section 25 of the Theft Act 1968 (going equipped for stealing, etc.) committed with reference to the theft or taking of motor vehicles.	Discretionary.		

Schedule 3 to the Road Traffic Offenders Act 1988

SCHEDULE 3
FIXED PENALTY OFFENCES

(1) Provision creating offence	(2) General nature of offence
Offences under the Parks Regulation (Amendment) Act 1926	
Section 2(1).	Breach of parks regulations but only where the offence is committed in relation to regulation 4(27) (driving or riding a trade vehicle), 4(28) (exceeding speed limit) or 4(30) (unauthorised waiting by a vehicle or leaving a vehicle unattended) of the Royal and other Parks and Gardens Regulations 1977.
Offences under the Highways Act 1835 and the Roads (Scotland) Act 1984	
Section 72 of the Highways Act 1835.	Driving on the footway. Cycling on the footway.
Section 129(5) of the Roads (Scotland) Act 1984.	Driving on the footway.
Offence under the Greater London Council (General Powers) Act 1974 (c.xxiv)	
Section 15 of the Greater London Council (General Powers) Act 1974.	Parking vehicles on footways, verges, etc.
Offence under the Highways Act 1980	
Section 137 of the Highways Act 1980.	Obstructing a highway, but only where the offence is committed in respect of a vehicle.
Offences under the Road Traffic Regulation Act 1984	
RTRA section 5(1)	Using a vehicle in contravention of a traffic regulation order outside Greater London.
RTRA section 8(1)	Breach of traffic regulation order in Greater London.
RTRA section 11	Breach of experimental traffic order.
RTRA section 13	Breach of experimental traffic scheme regulations in Greater London.
RTRA section 16(1)	Using a vehicle in contravention of temporary prohibition or restriction of traffic in case of execution of works, etc.
RTRA section 17(4)	Wrongful use of special road.
RTRA section 18(3)	Using a vehicle in contravention of provision for one-way traffic on trunk road.
RTRA section 20(5)	Driving a vehicle in contravention of order prohibiting or restricting driving vehicles on certain classes of roads.

(1) Provision creating offence	(2) General nature of offence
	Offences under the Road Traffic Regulation Act 1984—continued
RTRA section 25(5)	Breach of pedestrian crossing regulations, except an offence in respect of a moving motor vehicle.
RTRA section 29(3)	Using a vehicle in contravention of a street playground order.
RTRA section 35A(1)	Breach of an order regulating the use, etc., of a parking place provided by a local authority, but only where the offence is committed in relation to a parking place provided on a road.
RTRA section 47(1)	Breach of a provision of a parking place designation order and other offences committed in relation to a parking place designated by such an order, except any offence of failing to pay an excess charge within the meaning of section 46.
RTRA section 53(5)	Using vehicle in contravention of any provision of a parking place designation order having effect by virtue of section 53(1)(a) (inclusion of certain traffic regulation provisions).
RTRA section 53(6)	Breach of a provision of a parking place designation order having effect by virtue of section 53(1)(b) (use of any part of a road for parking without charge).
RTRA section 88(7)	Driving a motor vehicle in contravention of an order imposing a minimum speed limit under section 88(1)(b).
RTRA section 89(1)	Speeding offences under RTRA and other Acts.
	Offences under the Road Traffic Act 1988
RTA section 14	Breach of regulations requiring wearing of seat belts.
RTA section 15(2)	Breach of restriction on carrying children in the front of vehicles.
RTA section 15(4)	Breach of restriction on carrying children in the rear of vehicles.
RTA section 16	Breach of regulations relating to protective headgear for motor cycle drivers and passengers.
RTA section 19	Parking a heavy commercial vehicle on verge or footway.
RTA section 22	Leaving vehicle in dangerous position.
RTA section 23	Unlawful carrying of passengers on motor cycles.
RTA section 24	Carrying more than one person on a pedal cycle.
RTA section 34	Driving mechanically propelled vehicle elsewhere than on a road.
RTA section 35	Failure to comply with traffic directions.
RTA section 36	Failure to comply with traffic signs.
RTA section 40A	Using vehicle in dangerous condition etc.
RTA section 41A	Breach of requirement as to brakes, steering-gear or tyres.
RTA section 41B	Breach of requirement as to weight: goods and passenger vehicles.
RTA section 42	Breach of other construction and use requirements.
RTA section 47	Using, etc., vehicle without required test certificate being in force.
RTA section 87(1)	Driving vehicle otherwise than in accordance with requisite licence.
RTA section 143	Using motor vehicle while uninsured or unsecured against third party risks.
RTA section 163	Failure to stop vehicle on being so required by constable in uniform.
RTA section 172	Failure of person keeping vehicle and others to give the police information as to identity of driver, etc., in the case of certain offences.
	Offences under the Vehicle Excise and Registration Act 1994
Section 33 of the Vehicle Excise and Registration Act 1994.	Using or keeping a vehicle on a public road without licence being exhibited in manner prescribed by regulations.
Section 42 of that Act.	Driving or keeping a vehicle without required registration mark.
Section 43 of that Act.	Driving or keeping a vehicle with registration mark obscured etc.
Section 59 of that Act.	Failure to fix prescribed registration mark to a vehicle in accordance with regulations made under section 23(4)(a) of that Act.

Appendix 4

CPS Driving Offences Charging Standard

INDEX

DRIVING OFFENCES CHARGING STANDARD AGREED BY THE POLICE AND CROWN PROSECUTION SERVICE

1 Charging Standard—Purpose

1.1 The purpose of joint charging standards is to make sure that the most appropriate charge is selected, in the light of the evidence which can be proved, at the earliest possible opportunity. This will help the police and Crown Prosecutors in preparing the case. Adoption of this joint standard should lead to a reduction in the number of times charges have to be amended which in turn should lead to an increase in efficiency and a reduction in avoidable extra work for the police and the Crown Prosecution Service.

1.2 This joint charging standard offers guidance to police officers who have responsibility for charging and to Crown Prosecutors on the most appropriate charge to be preferred in cases relating to driving offences. The guidance:

- **should not be used** in the determination of any **pre-charge** decision, such as the decision to arrest;
- **does not** override any guidance issued on the use of appropriate alternative forms of disposal **short of charge**, such as cautioning;

- **does not** override the principles set out in the Code for Crown Prosecutors;
- **does not** override the need for consideration to be given in every case as to whether a charge/prosecution is in the public interest;
- **does not** remove the need for each case to be considered on its individual merits or fetter the discretion of the police to charge and the CPS to prosecute the most appropriate offence depending on the particular facts of the case in question.

2 Introduction

2.1 The purpose of road traffic legislation is to promote road safety and to protect the public. The principal driving offences are contained in the Road Traffic Act 1988 ('RTA 1988'). This joint standard gives guidance about the charge which should be preferred if the criteria set out in the Code for Crown Prosecutors are met.

2.2 This standard covers the following offences:

- careless driving or inconsiderate driving—section 3 RTA 1988;
- dangerous driving—section 2 RTA 1988;
- causing death by careless driving when under the influence of drink or drugs—section 3A RTA 1988;
- causing death by dangerous driving—section 1 RTA 1988;
- manslaughter—contrary to common law;
- causing bodily harm by wanton or furious driving, etc.—section 35 of the Offences Against the Person Act 1861.

3 General Principles: Charging Practice

3.1 You should always have in mind the following general principles when selecting the appropriate charge(s):

(i) the charge(s) should accurately reflect the extent of the defendant's alleged involvement and responsibility thereby allowing the courts the discretion to sentence appropriately;

(ii) the choice of charges should ensure the clear and simple presentation of the case particularly where there is more than one defendant;

(iii) it is wrong to encourage a defendant to plead guilty to a few charges by selecting more charges than are necessary;

(iv) it is wrong to select a more serious charge which is not supported by the evidence in order to encourage a plea of guilty to a lesser allegation.

4 General comments about driving offences

4.1 The manner of the driving must be considered objectively. In practice, the difference between the two types of bad driving will depend on the degree to which the driving falls below the minimum acceptable standard. If the manner of the driving is **below** that which is expected, the appropriate charge will be careless driving; if the manner of the driving is **far below** that which is expected, the appropriate charge will be dangerous driving. There is no statutory guidance about what behaviour constitutes a manner of driving which is 'below' and 'far below' the required standard.

4.2 The purpose of this charging standard is to make sure that once a decision to prosecute has been taken, police officers and prosecutors select the most appropriate charge where there is a choice of two or more charges. The following factors are not relevant when deciding whether an act of driving is careless or dangerous:

- the injury or death of one or more persons involved in a road traffic accident, except where Parliament has made specific provision for the death to be reflected in the charge. Importantly, injury or death does not, by itself, turn an accident into careless driving or turn careless driving into dangerous driving;

- the age or experience of the driver;
- the commission of other driving offences at the same time (such as driving whilst disqualified or driving without a certificate of insurance or a driving licence);
- the fact that the defendant has previous convictions for road traffic offences;
- the disability of a driver caused by mental illness or by physical injury or illness, except where the disability adversely affected the manner of the driving.

4.3 There is no clear cut dividing line between acts of careless driving and acts of dangerous driving. True momentary inattention will not usually have very serious adverse effects and therefore does not usually warrant criminal proceedings. Something more than momentary inattention (which may have minimal or serious results) is generally careless driving. Substantial/gross/total inattention (which may have minimal or serious results) is generally dangerous driving, even though it may take place over a period of a few seconds. The factual examples set out in this standard are merely indicative of the sort of behaviour which may merit prosecution under either section 2 or section 3 RTA 1988.

4.4 It is important to put the facts of the case in context. Although the test is objective, the manner of the driving must be seen in the context of the circumstances in which the driving took place. Behaviour which may not be criminal in certain conditions may merit proceedings in other conditions, for example, a safe lane change in slow moving traffic may become unsafe on a motorway where speeds are faster, there is less time to react and the consequences of any accident are likely to be more serious. Similarly, behaviour which might merit proceedings under section 3 in certain conditions may merit a prosecution under section 2, for example, if there is poor visibility; increased volume of traffic; adverse weather conditions; or difficult geography, such as blind corners.

Driving in emergency situations

4.5 When a member of the emergency services commits an offence while responding to an emergency call discretion should be used in deciding whether or not a prosecution is needed. Generally, a prosecution is unlikely to be appropriate in cases of genuine emergency unless the driving is dangerous or indicates a high degree of blameworthiness. For example, a prosecution of a driver who caused a minor accident while responding to an urgent, life threatening, emergency may not be appropriate; but a prosecution may be appropriate when a serious accident is caused by an over-enthusiastic driver responding to a less urgent emergency call in which life is not threatened. In each case it is necessary to weigh all the circumstances of the case, particularly the nature of the emergency known to, or reasonably perceived by, the driver and the nature of the driving.

4.6 There will be cases when persons who are not members of the emergency services drive in an emergency situation. Examples include doctors who receive an urgent call for assistance and a driver taking a sick child to hospital. As with members of the emergency services, all the circumstances of the case must be weighed, particularly the nature of the emergency known to, or reasonably perceived by, the driver and the nature of the driving.

Driving and alcohol/drugs

4.7 The road traffic legislation treats the consumption of alcohol and drugs alike. The following principles apply to driving affected by the consumption of alcohol or drugs, though the case law, and the following paragraphs, focus on alcohol.

4.8 Assessing the relevance of the consumption of alcohol is a difficult area. The leading authority is *R* v *Woodward (Terence)* [1995] 1 WLR 375 (Court of Appeal). The following general principles come from that case:

- the mere fact that the driver has had drink is not of itself relevant to or admissible on the question of whether his driving is careless or dangerous;

- for such evidence to be admissible, it must tend to show that the amount of drink taken is such as would adversely affect a reasonable driver, or alternatively, that the accused was in fact adversely affected.

4.9 In practice, however, there will need to be some further evidence to show that the manner of the driving fell below or far below that which is to be expected in order to justify proceedings under section 3 or section 2 respectively.

5 Careless driving—section 3 RTA 1988

5.1 The offence of driving without due care and attention is committed when the driving falls below the standard expected of a reasonable, prudent and competent driver in all the circumstances of the case. It is a summary only offence carrying a level 4 fine (£2,500), discretionary disqualification for any period and/or until a driving test has been passed. The court must endorse the driver's licence with 3–9 penalty points unless there are special reasons not to do so.

5.2 The test of whether the standard of driving has fallen below the required standard is an objective one. It applies both when the manner of driving in question is deliberate and when the manner of driving occurs as a result of an error of judgement or simply as a result of incompetence or inexperience.

5.3 Section 38(7) RTA 1988 states that failure on the part of a person to observe a provision of the Highway Code shall not of itself render that person liable to criminal proceedings, but a failure, particularly a serious one, may constitute evidence of careless or dangerous driving.

5.4 In general, prosecution for careless driving will be appropriate when the manner of the driving demonstrates a serious miscalculation or a disregard for road safety, taking into account all the circumstances including road, traffic and/or weather conditions.

5.5 There will be rare occasions where an accident occurs and yet there is no evidence of any mechanical defect, illness of the driver or other explanation to account for why the accident happened. In these cases, a charge of careless driving may be appropriate. The prosecution can provide evidence to the court about the accident on the basis that in the absence of any explanation—such as the ones identified above—it is inevitable that the defendant must have been driving below the standard expected of a reasonable, prudent and competent driver, since otherwise the accident would not have happened.

5.6 The following are examples of driving which may support an allegation of careless driving:

1 acts of driving caused by more than momentary inattention and where the safety of road users is affected, such as:
   - overtaking on the inside;
   - driving inappropriately close to another vehicle;
   - driving through a red light;
   - emerging from a side road into the path of another vehicle;
   - turning into a minor road and colliding with a pedestrian.

2 conduct which clearly caused the driver not to be in a position to respond in the event of an emergency on the road, for example:
   - using a hand held mobile telephone while the vehicle is moving, especially when at speed;
   - tuning a car radio;
   - reading a newspaper/map;
   - selecting and lighting a cigarette/cigar/pipe;
   - talking to and looking at a passenger which causes the driver more than momentary inattention;
   - leg and/or arm in plaster;
   - fatigue/nodding off.

The above examples explain the driver's conduct rather than demonstrate a course of driving which necessarily falls below the objective standard of the driving itself. For example, they may explain why the driver veered across carriageways, passed through a red light or otherwise caused a danger to other road users. In these cases, it is necessary to go beyond the explanation for the driving and consider whether the particular facts of the case warrant a charge of careless or dangerous driving. The reason for the driver's behaviour is not relevant to the choice of charge: it is the acts of driving which determine whether the driver has fallen below (careless driving) or far below (dangerous driving) the standard required.

These examples are placed here because *usually* when this conduct occurs the appropriate charge will be section 3. But police officers and prosecutors must always consider the manner of the driving in the context of the other facts in the case to decide the most appropriate way forward.

5.7 In deciding whether a charge of careless driving is appropriate, you will want to consider whether the act of driving concerned was the result of either momentary inattention or an isolated misjudgment, or something more serious. A moment's inattention which causes the manner of the driving to fall below the objective standard required of the reasonable, prudent and competent driver need not, of itself, lead to a prosecution. It is acts caused by more than momentary inattention—especially where the manner of the driving adversely affects the safety of other road users—which will normally result in a charge of careless driving.

5.8 In cases where there has been an accident and the evidence suggests that more than one driver may have been at fault, it will be necessary to establish that there is independent evidence against each driver before charging any individual driver, or that the facts speak so strongly for themselves in relation to any individual driver that the only conclusion possible to draw is that he departed from what a reasonable, prudent and competent driver would have done in the circumstances.

5.9 The public interest in favour of a prosecution is proportionate to the degree of blameworthiness: the greater the blameworthiness, the greater the public interest in favour of prosecuting. In addition, the public interest will favour prosecuting in cases when the court may wish to make an order under section 36 of the Road Traffic Offenders Act 1988, disqualifying the driver until he passes a driving test; or where it appears that the court ought to notify the Secretary of State that the driver may be suffering from any relevant disability within the meaning of section 22 Road Traffic Offenders Act 1988.

5.10 However, the public interest does not call for a prosecution in every case where there is a realistic prospect of conviction for careless driving: prosecution for an act of slight carelessness is unlikely to have a deterrent effect; and it is not the function of the prosecution to conduct proceedings merely to settle questions of liability for the benefit of insurance companies.

5.11 The public interest will tend to be against a prosecution for careless driving where:

- the incident is of a type such as frequently occurs at parking places, roundabouts, junctions or in traffic queues, involving minimal carelessness such as momentary inattention or a minor error of judgement;
- only the person at fault suffered injury and damage, if any, was mainly restricted to the vehicle or property owned by that person.

5.12 In addition, there is often an overlap between careless driving and some other offences such as drivingwith excess alcohol, regulatory offences, offences of strict liability, or offences under the Road Vehicles (Construction and Use) Regulations 1986. The merits of many individual cases can be adequately met by charging the specific statutory or regulatory offence which Parliament made available, subject to paragraphs 5.13 and 5.14 below.

5.13 Sometimes, there will be evidence of a course of conduct which involves the commission of a number of different statutory or regulatory offences, or the commission of the same statutory or regulatory offence on a number of occasions which are very close in time with one another. For example, a driver may drive through a red traffic light, ignore a pelican crossing and fail to give way at a junction within what might reasonably be described as the same course of driving. Alternatively, a driver may drive through two or more sets of red traffic lights, one after the other, within what may reasonably be described as the same course of driving.

5.14 In these situations, it is not appropriate simply to charge a number of individual statutory or regulatory offences: the court needs to be made aware of the link between what might otherwise appear as isolated incidents, when in reality they form part of a more serious course of bad driving. This course of bad driving should be reflected in a more serious charge. Where this type of situation arises, the manner of the driving has, in reality, fallen far below that expected of a competent and careful driver because of the driver's systematic failure to pay heed to the relevant traffic directions. Accordingly, consideration should be given to prosecuting the driver under section 2 of the Act: see paragraph 7.

6 Driving without reasonable consideration—section 3 RTA 1988

6.1 The offence of driving without reasonable consideration is committed when a vehicle is driven on a road or other public place as a result of which other persons using the road or place are inconvenienced. It is a summary only offence carrying a level 4 fine (£2,500), discretionary disqualification for any period and/or until a driving test has been passed. The court must endorse the driver's licence with 3–9 penalty points unless there are special reasons not to do so.

6.2 The accused must be shown:

- to have fallen below the standard of a reasonable, prudent and competent driver in the circumstances of the case; **and**
- to have done so without reasonable consideration for others.

6.3 The difference between the two offences under section 3 is that in cases of careless driving the prosecution need not show that any other person was inconvenienced. In cases of inconsiderate driving, there must be evidence that some other user of the road or public place was inconvenienced.

6.4 An allegation of inconsiderate driving is appropriate when the driving amounts to a clear act of selfishness, impatience or aggressiveness. There must, however, also be some inconvenience to other road users, for example, forcing other drivers to move over and/or brake as a consequence. Examples of conduct appropriate for a charge of driving without reasonable consideration are:

- flashing of lights to *force* other drivers in front to give way;
- misuse of any lane to avoid queuing or gain some other advantage over other drivers;
- unnecessarily remaining in an overtaking lane;
- unnecessarily slow driving or braking without good cause;
- driving with undipped headlights which dazzle oncoming drivers;
- driving through a puddle causing pedestrians to be splashed.

6.5 A person who drives without reasonable consideration for other road users can be convicted of driving without due care and attention although the reverse does not apply.

7 Dangerous Driving—section 2 of the Act

7.1 A person drives dangerously when:

- the way he drives falls **far below** what would be expected of a competent and careful driver; **and**
- it would be obvious to a competent and careful driver that driving in that way would be dangerous.

7.2 Both parts of the definition must be satisfied for the driving to be 'dangerous' within the Act. Dangerous driving is an either way offence. In the magistrates' courts the maximum penalty is a level 5 fine (£5,000), and/or six months imprisonment; in the Crown Court, the maximum penalty is 2 years imprisonment and/or an unlimited fine. In both instances, the court must disqualify the driver from driving for at least one year and must endorse the driver's licence with 3–11 penalty points unless in either case there are special reasons not to do so.

7.3 The test of whether a driver has fallen far below the required standard is an objective one. It applies both when the manner of driving in question is deliberate and when the manner of driving occurs as a result of an error of judgement or simply as a result of incompetence or inexperience.

7.4 There is no statutory definition of what is meant by 'far below', but 'dangerous' must refer to danger either of injury to any person or of serious damage to property: s. 2A(3) of the Act. Additionally, s. 2A(2) of the Act provides that a person is to be regarded as driving dangerously if it would be obvious to a competent and careful driver that driving the vehicle in its current state would be dangerous. When considering the 'state' of the vehicle, regard may be had to anything attached to or carried by the vehicle: section 2A(4) of the Act. Therefore, you must consider whether the vehicle should have been driven at all, as well as how it was driven.

7.5 The standard of driving must be objectively assessed. It is not necessary to consider what the driver thought about the possible consequences of his actions. What must be considered is whether or not a competent and careful driver would have observed, appreciated and guarded against obvious and material dangers.

7.6 In deciding whether a charge of dangerous driving is appropriate, you will want to consider whether the act of driving concerned was undertaken deliberately and/or repeatedly. Although the test of dangerousness is an objective one, deliberate or repeated disregard, for example, of traffic directions (be they 'stop' or 'give way' signs or traffic lights) may be evidence that the manner of the accused's driving has fallen far below the standard required, thereby making a charge of dangerous driving appropriate.

7.7 In addition, the following are examples of driving which may support an allegation of dangerous driving:

- racing or competitive driving;
- prolonged, persistent or deliberate bad driving;
- speed which is highly inappropriate for the prevailing road or traffic conditions;
- aggressive or intimidatory driving, such as sudden lane changes, cutting into a line of vehicles or driving much too close to the vehicle in front, especially when the purpose is to cause the other vehicle to pull to one side to allow the accused to overtake;
- disregard of traffic lights and other road signs, which, on an objective analysis, would appear to be deliberate;
- failure to pay proper attention, amounting to something significantly more than a momentary lapse;
- overtaking which could not have been carried out with safety;
- driving a vehicle with a load which presents a danger to other road users.

8 Causing death by careless driving when under the influence of drink or drugs—section 3A RTA 1988

8.1 This offence is committed when:

- the driving was without due care and attention or without reasonable consideration for other road users; **and**
- the driving has caused the death of another person; **and**
- the driver is either unfit through drink or drugs, or the alcohol concentration is over the prescribed limit, or there has been a failure to provide a specimen in pursuance of the RTA 1988.

8.2 It is an offence triable only on indictment and carries a maximum penalty of 10 years imprisonment and/or an unlimited fine and an obligatory disqualification for at least 2 years (3 years if there is a previous relevant conviction). The driver's licence must be endorsed with 3–11 penalty points.

8.3 The examples given in paragraph 5 of careless driving apply to this offence; in the context of section 3A, less serious examples of careless driving (which may not of themselves require a prosecution under section 3 alone) may also merit proceedings under section 3A.

8.4 The accused's driving must have been a cause of the death but need not be the sole one.

8.5 Proper procedures have to have been adopted in the requesting and/or obtaining of any sample of breath, blood or urine. In cases where the procedures are flawed, there is a risk that the evidence may be excluded. Where this is possible, careful consideration must be given to whether the remaining evidence will support an alternative allegation of causing death by careless driving while unfit to drive through drink/drugs, in which case, evidence other than that from an intoximeter machine can be relied upon to demonstrate the defendant's unfitness to drive.

8.6 It is not necessary to add a further charge relating to drink/driving when the defendant is charged with a section 3A offence, because a guilty verdict to the relevant drink/drive offence can be returned by the jury under the statutory provisions: see paragraph 14.

9 Causing death by dangerous driving—section 1 RTA 1988

9.1 This offence is committed when:

- the driving of the accused was a cause of the death of another person **and**
- the driving was dangerous within the meaning of section 2A of the Act (see paragraph 7.3 of this standard).

9.2 The offence is triable only on indictment and carries a maximum penalty of 10 years imprisonment and/or an unlimited fine with an obligatory disqualification for a minimum of 2 years. The driver's licence must be endorsed with 3–11 penalty points.

9.3 The accused's driving must have been a cause of the death but need not be the sole one.

9.4 The examples given in paragraph 7 of dangerous driving apply to this offence.

9.5 Where a section 1 offence can be proved and there is sufficient evidence of a section 4, 5 or 7 offence, the appropriate summary offence should be charged and adjourned sine die pending the outcome of the section 1 offence—these offences cannot be committed to the Crown Court under section 41(1) Criminal Justice Act 1988. If the defendant is convicted of the section 1 offence, the court will often make it clear that the sentence imposed reflects the element of drink/driving, in which case the summary offence should not subsequently be pursued. Where the defendant is acquitted of the section 1 offence (or is convicted but it is clear the court has not taken the element of drink/driving into account) prosecutors should consider re-activating the drink/drive offence.

10 Relationship between section 1 and section 3A RTA 1988

10.1 Offences under section 1 and section 3A carry the same maximum penalty, so the choice of charge will not inhibit the court's sentencing powers. The courts have made it clear that for sentencing purposes the two offences are to be regarded on an equal basis. (*Attorney-General's Reference (No. 49 of 1994) R* v *Brown* [1995] Crim LR 437; *R* v *Locke* [1995] Crim LR 438.)

10.2 The court will sentence an offender in proportion to his criminality. The consumption of alcohol is an aggravating feature increasing the criminality of the offender and therefore the sentence passed. The consumption of alcohol is an aggravating feature within the definition of section 3A. The consumption of alcohol is not part of the definition of section 1 but may be treated as an aggravating feature in appropriate cases.

10.3 Where a section 1 offence can be proved, it should be charged. However, you may on occasions have to decide which is the more appropriate charge: section 1 or section 3A. This will almost always occur when the manner of the driving is on the borderline between careless and dangerous. The prosecution is likely to be put to election if the two offences are charged in the alternative. In borderline cases, section 3A should be chosen provided all the other elements of that offence can be proved. The prospects of a conviction will be greater and the court's sentencing power remains unaffected.

11 Manslaughter—contrary to common law

11.1 Manslaughter is committed when the driver, in breach of a duty of care, is criminally negligent and causes the death of the victim.

11.2 The offence is triable only on indictment and carries a maximum sentence of life imprisonment and/or an unlimited fine. The driver must be disqualified for at least two years and there is a compulsory re-test. The driver's licence must be endorsed with 3–11 penalty points.

11.3 The driver must be shown to have been in breach of a duty of care towards the person who died. The ordinary principles of the law of negligence apply to ascertain whether there is such a duty. There is a general duty of care on all persons not to do acts imperilling the lives of others. To show a breach of a duty of care will require proof that the driving:

- fell far below the minimum acceptable standard of driving; **and**
- involved a risk of death; **and**
- was so bad in all the circumstances as, in the opinion of the jury, to amount to a crime: *R* v *Adomako* [1994] 3 All ER 79.

11.4 The examples of driving which fall far below the minimum acceptable standard of driving set out in paragraph 7.7 apply here as well.

11.5 This charge will very rarely be appropriate in road traffic fatality cases because of the existence of the statutory offences.

11.6 Manslaughter should be considered when a vehicle has been used as an instrument of attack (but where the necessary intent for murder is absent) or to cause fright and death results.

11.7 Manslaughter should also be considered where the driving has occurred other than on a road or other public place, or when the vehicle driven was not mechanically propelled, and death has been caused. In these cases the statutory offences do not apply.

12 Causing bodily harm by wanton and furious driving—section 35 Offences Against the Person Act 1861

12.1 It is an offence for any person in charge of a vehicle:

- to cause *or cause to be done* bodily harm to any person; by
- wanton or furious driving, or other wilful misconduct, or by wilful neglect.

12.2 It is a offence triable only on indictment and carries a maximum penalty of a 2 years imprisonment and/or an unlimited fine.

12.3 This offence should be used rarely as it does not carry endorsement or disqualification. It should normally only be used on those occasions when it is not possible to prosecute for an offence under the road traffic legislation, for example:

- when the driving was not on a road or other public place;
- when the vehicle used is not a mechanically propelled vehicle within the RTA 1988;
- when the statutory notice of intended prosecution is a prerequisite to a prosecution and has not been given.

12.4 This offence is useful in cases when a victim suffers serious injury though there has been no direct contact between the victim and the vehicle. For example, when the driving caused the victim to take avoiding action and as a result of which sustained serious injury by, say, falling down a ditch.

12.5 When a vehicle has been deliberately used as a weapon and has caused injury, alternative charges of dangerous driving under s. 2 of the Act or s. 18 Offences Against the Person Act 1861 should be considered if all the elements of those offences can be proved.

13 Road Traffic Fatality Cases: 'Nearest and Dearest'

13.1 In addition to the public interest considerations set out in the Code for Crown Prosecutors, special considerations apply to cases when there is a family or other close personal relationship between the deceased and the accused driver. These are often referred to as 'nearest and dearest' cases. The considerations are unlikely to be relevant in any case where the evidence would support proceedings for manslaughter.

13.2 In each case, the particular circumstances and the nature of the relationship will have to be considered. The closer the relationship between the deceased and the accused driver, the more likely it will be that the guidance which follows will apply.

13.3 In cases of causing death by dangerous driving involving the death of a 'nearest and dearest', where there is evidence to suggest an aggravating feature which imperils other road users or that the accused is a continuing danger to other road users, the proper course will be to prosecute for dangerous driving (section 2). The focus of the case will then be imperilling of other road users.

13.4 Additionally, if the accused drove in such a way as to show serious disregard for the lives of the 'nearest and dearest' or other road users, notwithstanding that a 'nearest and dearest' has been killed, proceedings for causing death by dangerous driving should be considered.

13.5 However, in cases of causing death by dangerous driving involving the death of a 'nearest and dearest', where there is **no** evidence either of an aggravating feature imperilling other road users nor that the accused is a continuing danger to other road users, the proper course is not to prosecute.

13.6 In cases of causing death by careless driving while under the influence of drink etc. involving the death of a 'nearest and dearest', the proper course will be to prosecute for careless driving and the appropriate drink/driving offence.

13.7 In cases of careless driving which caused the death of a 'nearest and dearest' where there is evidence to suggest that the accused is a continuing danger to other road users, the proper course is to prosecute for careless driving (section 3).

13.8 However, in cases of careless driving which caused the death of a 'nearest and dearest' where there is **no** evidence that the accused is a continuing danger to other road users, the proper course is not to prosecute.

13.9 Evidence that an accused presents a continuing danger to other road users may be found in his/her previous convictions or medical condition. In such cases, the court may wish to make an order under section 36 of the Road Traffic Offenders Act 1988, disqualifying the driver until he passes a driving test; or when it appears that the court ought to notify the Secretary of State that the driver may be suffering from any relevant disability within the meaning of section 22 Road Traffic Offenders Act 1988.

13.10 If a person other than a 'nearest and dearest' is killed as a result of the dangerous driving, notwithstanding the fact that a near relative has also been killed, a charge for causing death by dangerous driving should normally follow. In order to present the case fully to the court a separate charge for the death of the close relative cannot, in these circumstances, be avoided.

14 Alternative Verdicts

14.1 In certain circumstances, it is possible for a jury to find the accused not guilty of the offence charged but guilty of some other alternative offence. The general provisions are contained in section 6(3) Criminal Law Act 1967 and are supplemented by other provisions which relate to specific offences.

14.2 Section 24 of the Road Traffic Offenders Act allows for the return of alternative verdicts where the allegations in the indictment amount to, or include, an allegation of an offence specified in the table set out in that section. The relevant statutory provisions are:

Offence charged	**Alternative verdicts**
Section 1: death by dangerous driving	Section 2: dangerous driving Section 3: careless, and inconsiderate, driving
Section 2: dangerous driving	Section 3: careless, and inconsiderate, driving
Section 3A: causing death by careless driving while under the influence of drink or drugs	Section 3: careless, and inconsiderate, driving and/or the relevant offence from: Section 4(1): driving whilst unfit Section 5(1)(a): driving with excess alcohol Section 7(6): failing to provide a specimen

14.3 Where the accused is charged with an offence under section 3A RTA 1988 he may not be convicted as an alternative with any offence of attempting to drive: section 24(2) Road Traffic Offenders Act 1988.

14.4 In the very rare cases when manslaughter is charged, it will normally be prudent to prefer an alternative charge for causing death by dangerous driving if the driving took place on a road or other public place. Further, when manslaughter is charged there should be no difficulty in also charging as an alternative a Section 3A offence if it is made out, although such a situation is most unlikely to arise.

14.5 It is essential however, that the charge which most suits the circumstances is always preferred. It will never be appropriate to charge a more serious offence in order to obtain a conviction (whether by plea or verdict) to a lesser offence.

Appendix 5

Road Vehicles Lighting Regulations 1989, Regulation 27

TABLE

(1) Item No.	(2) Type of lamp, hazard warning signal device or warning beacon	(3) Manner of use prohibited
1.	Headlamp	(a) Used so as to cause undue dazzle or discomfort to other persons using the road. (b) Used so as to be lit when a vehicle is parked.
2.	Front fog lamp	(a) Used so as to cause undue dazzle or discomfort to other persons using the road. (b) Used so as to be lit at any time other than in conditions of seriously reduced visibility. (c) Used so as to be lit when a vehicle is parked.
3.	Rear fog lamp	(a) Used so as to cause undue dazzle or discomfort to the driver of a following vehicle. (b) Used so as to be lit at any time other than in conditions of seriously reduced visibility. (c) Save in the case of an emergency vehicle, used so as to be lit when a vehicle is parked.
4.	Reversing lamp	Used so as to be lit except for the purpose of reversing the vehicle.
5.	Hazard warning signal device	Used other than— (i) to warn persons using the road of a temporary obstruction when the vehicle is at rest; or (ii) on a motorway or unrestricted dual-carriageway, to warn following drivers of a need to slow down due to a temporary obstruction ahead; or (iii) in the case of a bus, to summon assistance for the driver or any person acting as a conductor or inspector on the vehicle. or (iv) in the case of a bus to which prescribed signs are fitted as described in sub-paragraphs (a) and (b) of regulation 17A(1), when the vehicle is stationary and children under the age of 16 years are entering or leaving, or are about to enter or leave, or have just left the vehicle.

TABLE (*continued*)

(1) Item No.	(2) Type of lamp, hazard warning signal device or warning beacon	(3) Manner of use prohibited
6.	Warning beacon emitting blue light and special warning lamp	Used so as to be lit except— (i) at the scene of an emergency; or (ii) when it is necessary or desirable either to indicate to persons using the road the urgency of the purpose for which the vehicle is being used, or to warn persons of the presence of the vehicle or a hazard on the road.
7.	Warning beacon emitting amber light	Used so as to be lit except— (i) at the scene of an emergency; (ii) when it is necessary or desirable to warn persons of the presence of the vehicle; and (iii) in the case of a breakdown vehicle, while it is being used in connection with, and in the immediate vicinity of, an accident or breakdown, or while it is being used to draw a broken-down vehicle.
8.	Warning beacon emitting green light	Used so as to be lit except whilst occupied by a medical practitioner registered by the General Medical Council (whether with full, provisional or limited registration) and used for the purposes of an emergency.
9.	Warning beacon emitting yellow light	Used so as to be lit on a road.
10.	Work lamp	(a) Used so as to cause undue dazzle or discomfort to the driver of any vehicle. (b) Used so as to be lit except for the purpose of illuminating a working area, accident, breakdown or works in the vicinity of the vehicle.
11.	Any other lamp	Used so as to cause undue dazzle or discomfort to other persons using the road.

Appendix 6

Motor Vehicles (Driving Licences) Regulations 1999, Schedule 2 and Regulations 11 and 19

SCHEDULE 2

CATEGORIES AND SUB-CATEGORIES OF VEHICLES FOR LICENSING PURPOSES

PART 1

(1) Category or sub-category	(2) Classes of vehicle included	(3) Additional categories and sub-categories
A	Motor bicycles.	B1, K and P.
A1	A sub-category of category A comprising learner motor bicycles.	P.
B	Motor vehicles, other than vehicles included in category A, F, K or P, having a maximum authorised mass not exceeding 3.5 tonnes and not more than eight seats in addition to the driver's seat, including: (i) a combination of any such vehicle and a trailer where the trailer has a maximum authorised mass not exceeding 750 kilograms, and (ii) a combination of any such vehicle and a trailer where the maximum authorised mass of the combination does not exceed 3.5 tonnes and the maximum authorised mass of the trailer does not exceed the unladen weight of the tractor vehicle.	F, K and P.
B1	A sub-category of category B comprising motor vehicles having three or four wheels and an unladen weight not exceeding 550 kilograms.	K and P.
B + E	Combinations of a motor vehicle and trailer where the tractor vehicle is in category B but the combination does not fall within that category.	None.
C	Motor vehicles having a maximum authorised mass exceeding 3.5 tonnes, other than vehicles falling within category D, F, G or H, including such vehicle drawing a trailer having a maximum authorised mass not exceeding 750 kilograms.	None.
C1	A sub-category of category C comprising motor vehicles having a maximum authorised mass exceeding 3.5 tonnes but not exceeding 7.5 tonnes, including any such vehicle drawing a trailer having a maximum authorised mass not exceeding 750 kilograms.	None.
D	Motor vehicles constructed or adapted for the carriage of passengers having more than eight seats in addition to the driver's seat, including such vehicle drawing a trailer having a maximum authorised mass not exceeding 750 kilograms.	None.

PART 1 (*continued*)

(1) Category or sub-category	(2) Classes of vehicle included	(3) Additional categories and sub-categories
D1	A sub-category of category D comprising motor vehicles having more than eight but not more than 16 seats in addition to the driver's seat and including such vehicle drawing a trailer with a maximum authorised mass not exceeding 750 kilograms.	None.
C + E	Combination of a motor vehicle and trailer where the tractor vehicle is in category C but the combination does not fall within that category.	B + E.
C1 + E	A sub-category of category C + E comprising combinations of a motor vehicle and trailer where: (a) the tractor vehicle is in sub-category C1, (b) the maximum authorised mass of the trailer exceeds 750 kilograms but not the unladen weight of the tractor vehicle, and (c) the maximum authorised mass of the combination does not exceed 12 tonnes.	B + E.
D + E	Combinations of a motor vehicle and trailer where the tractor vehicle is in category D but the combination does not fall within that category.	B + E.
D1 + E	A sub-category of category D + E comprising combinations of a motor vehicle and trailer where: (a) the tractor vehicle is in sub-category D1, (b) the maximum authorised mass of the trailer exceeds 750 kilograms but not the unladen weight of the tractor vehicle, (c) the maximum authorised mass of the combination does not exceed 12 tonnes, and (d) the trailer is not used for the carriage of passengers.	B + E.
F	Agricultural or forestry tractors including any such vehicle drawing a trailer, but excluding any motor vehicle included in category H.	K.
G	Road rollers.	None.
H	Track-laying vehicles steered by their tracks.	None.
K	Mowing machines which do not fall within category A and vehicles controlled by a pedestrian.	None.
P	Mopeds.	None.
C1 + E (8.25 tonnes)	A sub-category of category C + E comprising combinations of a motor vehicle and trailer in sub-category C1 + E, the maximum authorised mass of which does not exceed 8.25 tonnes.	None.
D1 (not for hire or reward)	A sub-category of category D comprising motor vehicles in sub-category D1 driven otherwise than for hire or reward.	None.
D1 + E (not for hire or reward)	A sub-category of category D + E comprising motor vehicles in sub-category D1 + E driven otherwise than for hire or reward.	None.
L	Motor vehicles propelled by electrical power.	None.

11 Eligibility to apply for provisional licence

(1) Subject to the following provisions of this regulation, an applicant for a provisional licence authorising the driving of motor vehicles of a class included in a category or sub-category specified in column (1) of the table at the end of this regulation must hold a relevant full licence authorising the driving of motor vehicles of a class included in the category or sub-category specified in column (2) of the table in relation to the first category.

(2) Paragraph (1) shall not apply in the case of an applicant who is a full-time member of the armed forces of the Crown.

(3) For the purposes of paragraph (1), a licence authorising the driving only of vehicles in sub-categories D1 (not for hire or reward), D1 + E (not for hire or reward) and C1 + E (8.25 tonnes) shall not be treated as a licence authorising the driving of motor vehicles of a class included in sub-categories D1, D1 + E and C1 + E.

(4) In this regulation, 'relevant full licence' means a full licence granted under Part III of the Traffic Act, a full Northern Ireland licence, a full British external licence (other than a licence which is to be disregarded for the purposes of section 89(1)(d) of the Traffic Act by virtue of section 89(2)(c) of that Act), a full British Forces licence, an exchangeable licence or a Community licence.

TABLE

(1) Category or sub-category of licence applied for	(2) Category/sub-category of full licence required
B + E	B
C	B
C1	B
D	B
D1	B
C1 + E	C1
C + E	C
D1 + E	D1
D + E	D
G	B
H	B

19 Full licences not carrying provisional entitlement

(1) The application of sections 98(2) and 99A(5) of the Traffic Act is limited or excluded in accordance with the following paragraphs.

(2) Subject to paragraphs (3), (4), (5), (6), (11) and (12), the holder of a full licence which authorises the driving of motor vehicles of a class included in a category or sub-category specified in column (1) of the table at the end of this regulation may drive motor vehicles—

(a) of other classes included in that category or sub-category, and

(b) of a class included in each category or sub-category specified, in relation to that category or sub-category, in column (2) of the table, as if he were authorised by a provisional licence to do so.

(3) Section 98(2) shall not apply to a full licence if it authorises the driving only of motor vehicles adapted on account of a disability, whether pursuant to an application in that behalf made by the holder of the licence or pursuant to a notice served under section 92(5)(b) of the Traffic Act.

(4) In the case of a full licence which authorises the driving of a class of standard motor bicycles, other than bicycles included in sub-category A1, section 98(2) shall not apply so as to authorise the driving of a large motor bicycle by a person under the age of 21 before the expiration of the standard access period.

(5) In the case of a full licence which authorises the driving of motor bicycles of a class included in sub-category A1 section 98(2) shall not apply so as to authorise the driving of a large motor bicycle by a person under the age of 21.

(6) In the case of a full licence which authorises the driving of a class of vehicles included in category C or C + E, paragraph (2) applies subject to the provisions of regulation 54.

(7) Subject to paragraphs (8), (9), (10), (11) and (12), the holder of a Community licence to whom section 99A(5) of the Traffic Act applies and who is authorised to drive in Great Britain motor vehicles of a class included in a category or sub-category specified in column (1) of the Table at the end of this regulation may drive motor vehicles—

(a) of other classes included in that category or sub-category, and
(b) of a class included in each category or sub-category specified, in relation to that category or sub-category, in column (2) of the Table, as if he were authorised by a provisional licence to do so.

(8) Section 99A(5) shall not apply to a Community licence if it authorises the driving only of motor vehicles adapted on account of a disability.

(9) In the case of a Community licence which authorises the driving of a class of standard motor bicycle other than bicycles included in sub-category A1, section 99A(5) shall not apply so as to authorise the driving of a large motor bicycle by a person under the age of 21 before the expiration of the period of two years commencing on the date when that person passed a test for a licence authorising the driving of that class of standard motor bicycle (and in calculating the expiration of that period, any period during which that person has been disqualified for holding or obtaining a licence shall be disregarded).

(10) In the case of a Community licence which authorises the driving only of motor bicycles of a class included in sub-category A1 section 98(2) shall not apply so as to authorise the driving of a large motor bicycle by a person under the age of 21.

(11) Except to the extent provided in paragraph (12), section 98(2) shall not apply to a full licence, and section 99A(5) shall not apply to a Community licence, in so far as it authorises its holder to drive motor vehicles of any class included in category B + E, C + E, D + E or K or in sub-category B1 (invalid carriages), C1 or D1 (not for hire or reward).

(12) A person—
(a) who holds a full licence authorising the driving only of those classes of vehicle included in a category or sub-category specified in paragraph (11) which have automatic transmission (and are not otherwise adapted on account of a disability), or
(b) who holds a Community licence, to whom section 99A(5) of the Traffic Act applies and who is authorised to drive in Great Britain only those classes of vehicle included in a category or sub-category specified in paragraph (11) which have automatic transmission (and are not otherwise adapted on account of a disability), may drive motor vehicles of all other classes included in that category or sub-category which have manual transmission as if he were authorised by a provisional licence to do so.

TABLE

(1) Full licence held	(2) Provisional entitlement included
A1	A, B, F and K
A	B and F
B1	A, B and F
B	A, B + E, G and H
C	C1 + E, C + E
D1	D1 + E
D	D1 + E, D + E
F	B and P
G	H
H	G
P	A, B, F and K

Appendix 7

Motor Vehicles (Driving Licences) Regulations 1999, Schedule 4

SCHEDULE 4

DISTINGUISHING MARKS TO BE DISPLAYED ON A MOTOR VEHICLE BEING DRIVEN UNDER A PROVISIONAL LICENCE

PART 1

Diagram of distinguishing mark to be displayed on a motor vehicle in England, Wales or Scotland.

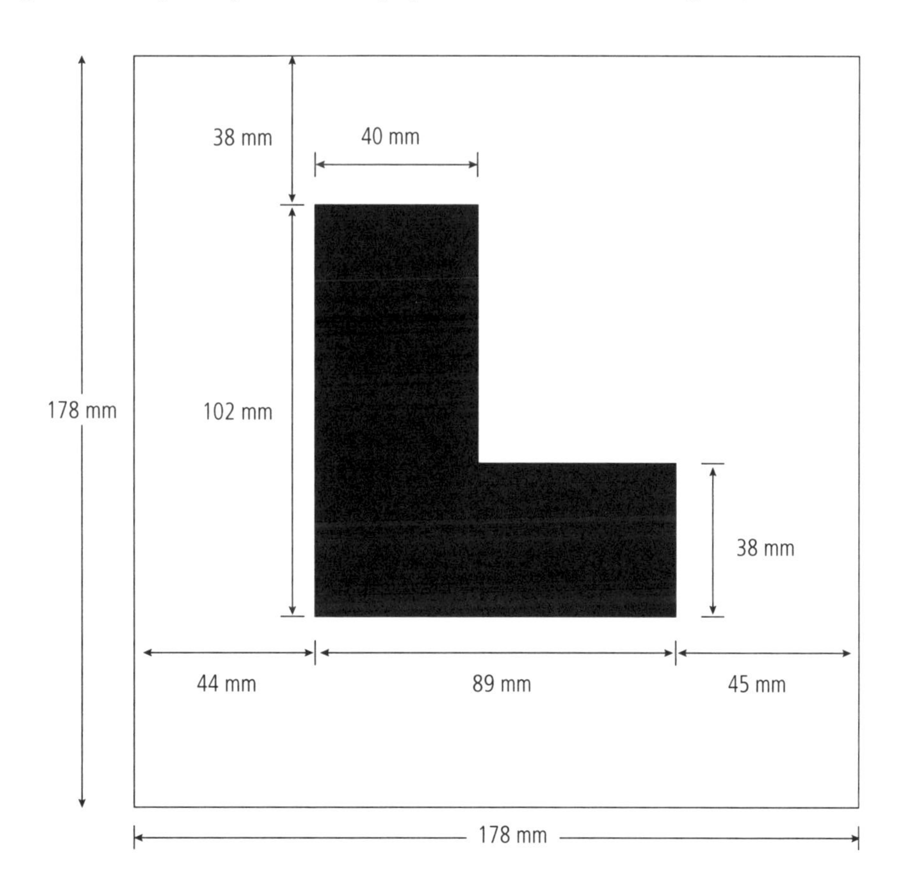

Red letter on white ground.
The corners of the ground can be rounded off.

PART 2

Diagram of optional distinguishing mark to be displayed on a motor vehicle in Wales if a mark in the form set out in Part 1 is not displayed.

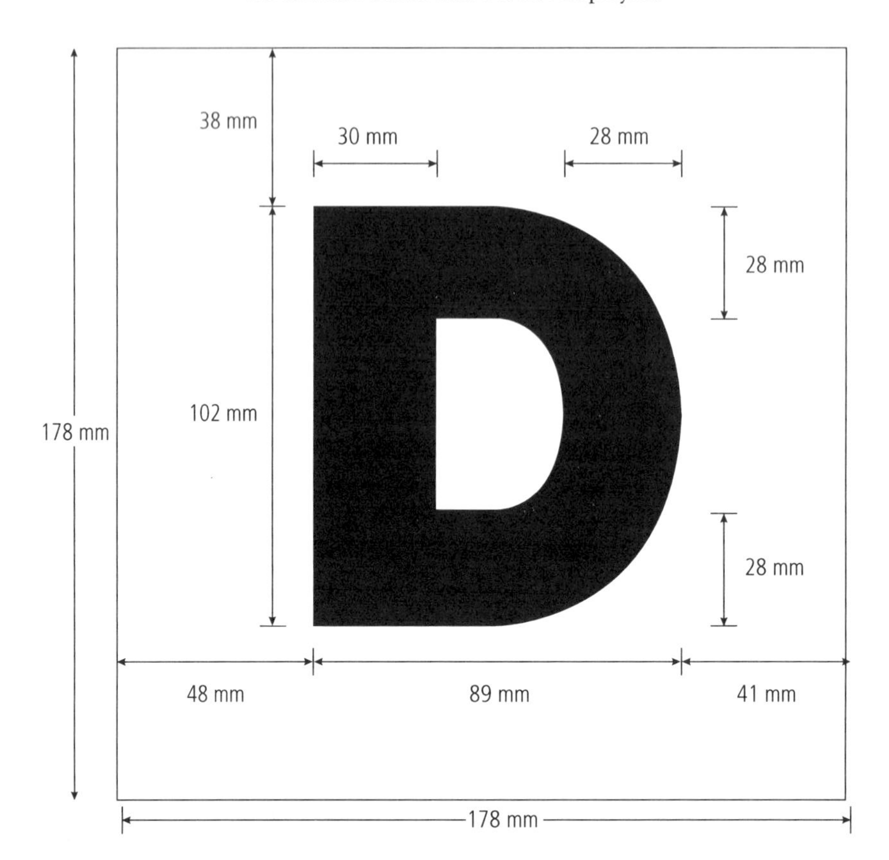

Red letter on white ground.
The corners of the ground can be rounded off.

Appendix 8

Screening breath tests

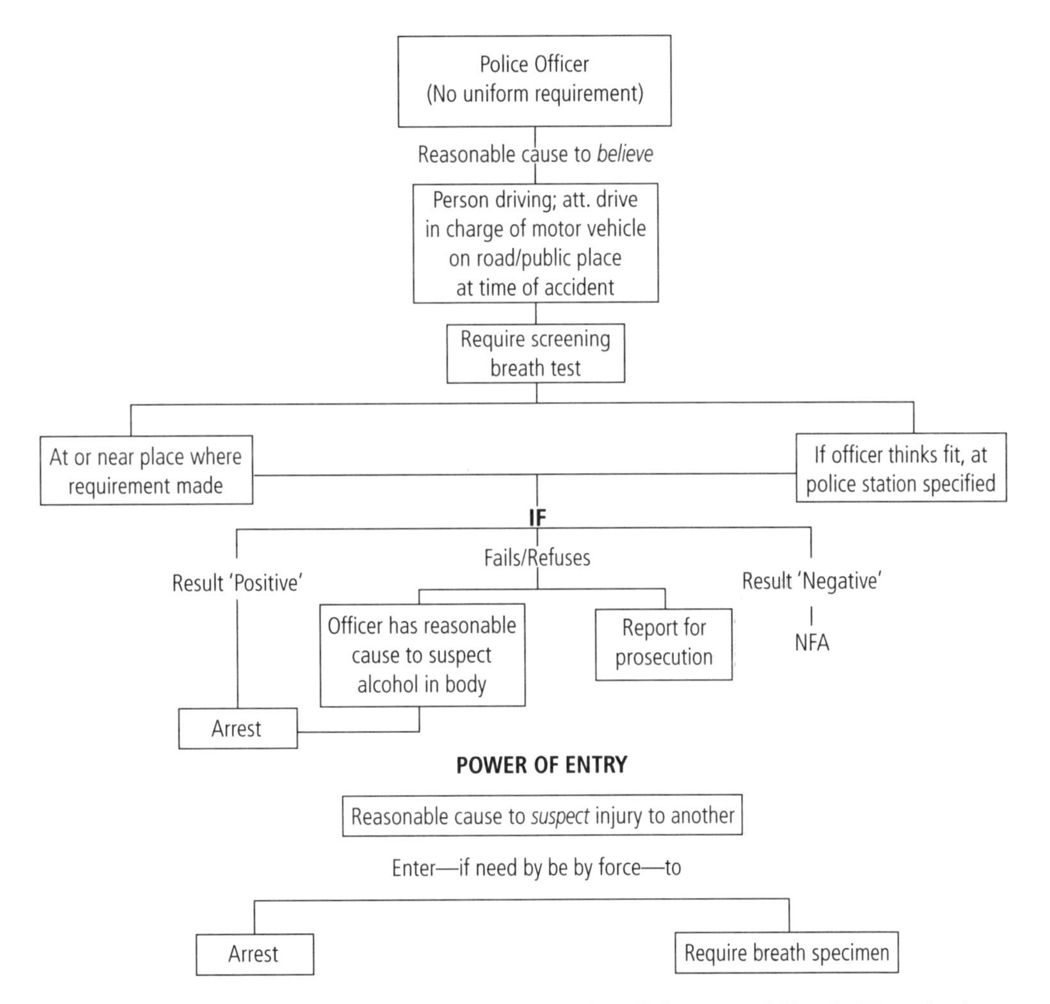

Diagram 1 Screening Breath Test Following Accident (Subject to s. 9 Hospital Procedure)

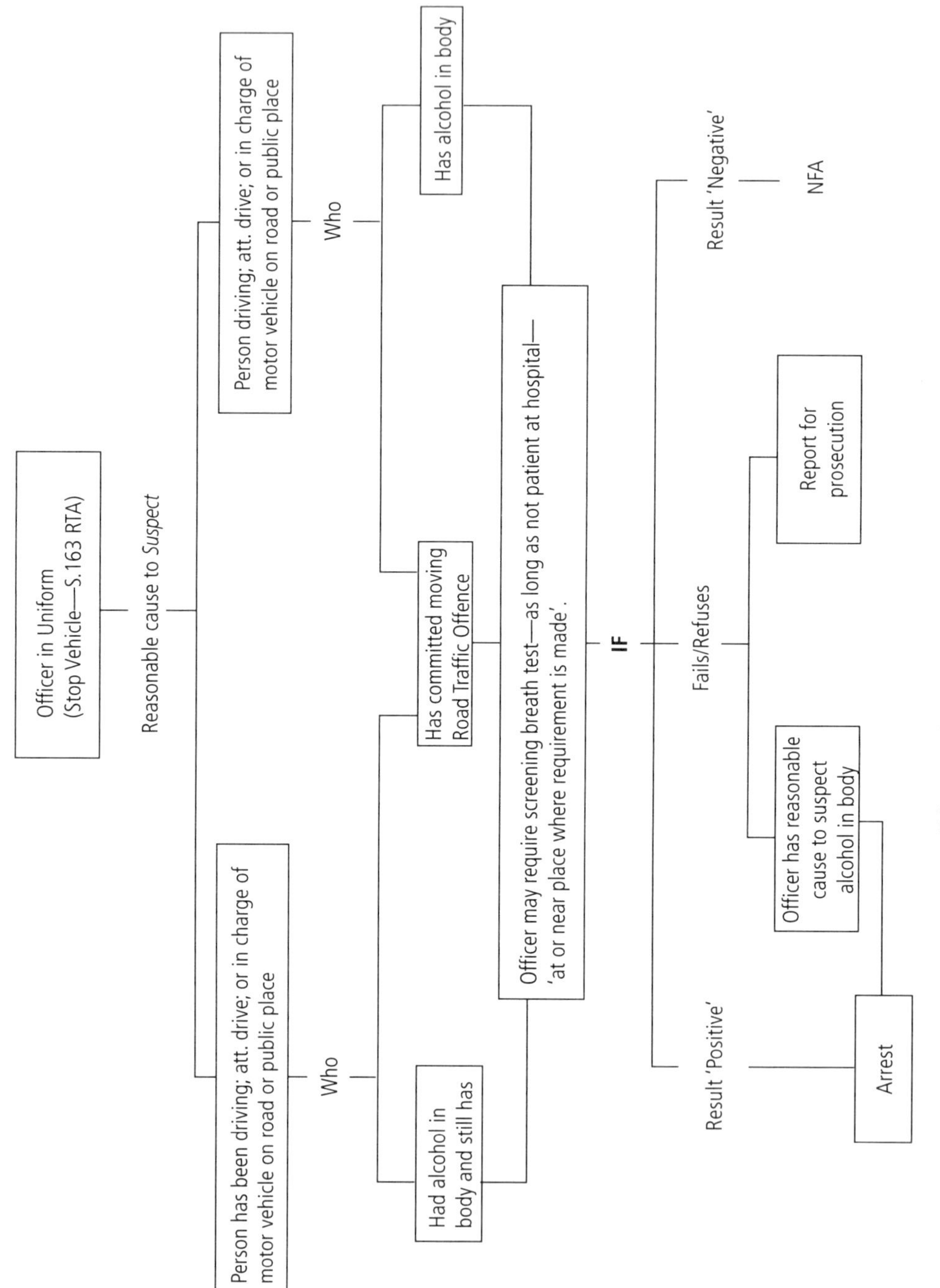

Diagram 2 Screening Breath Tests

Appendix 9

Fixed Penalty Procedure

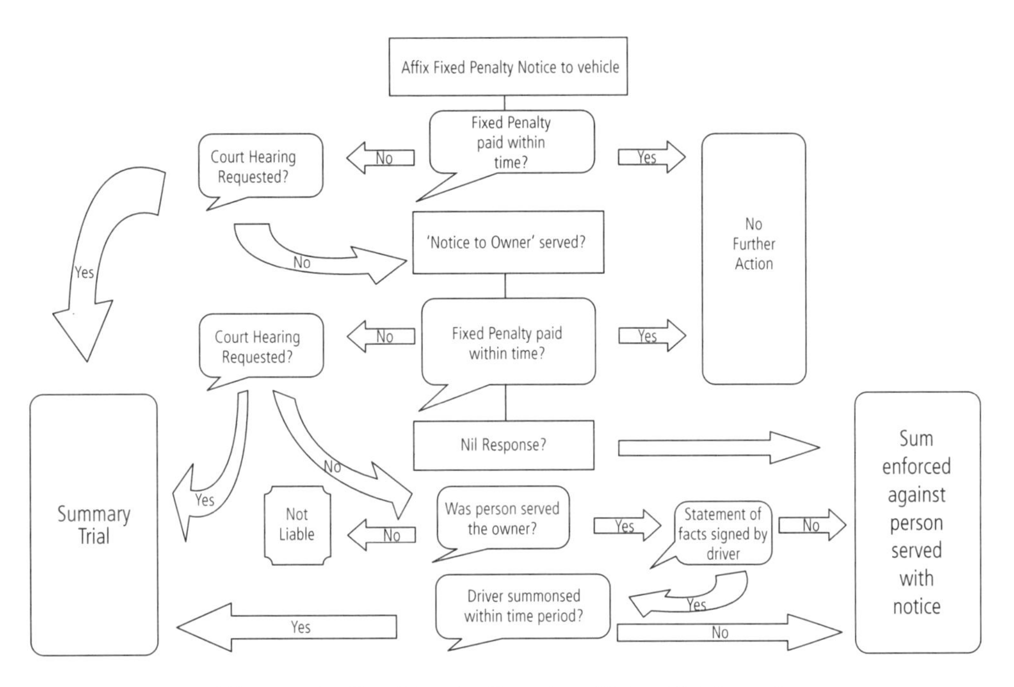

Diagram 1 Fixed Penalty Procedure (Driver Not Present)

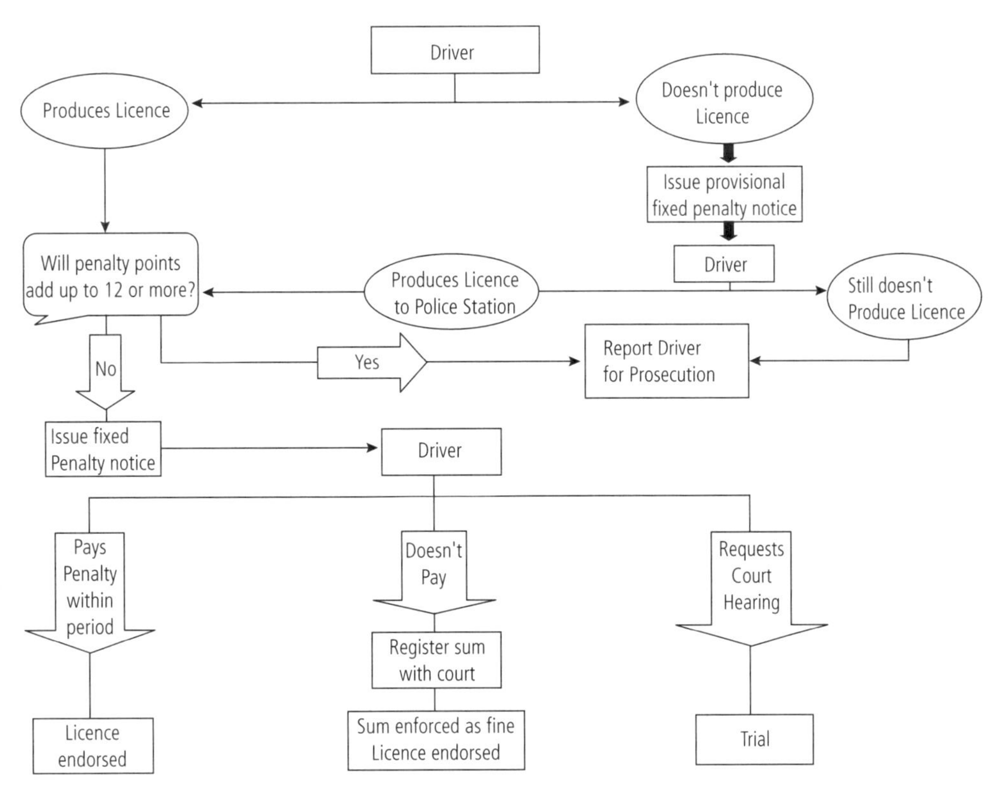

Diagram 2 Fixed Penalty Procedure—Endorsable (Driver Present)

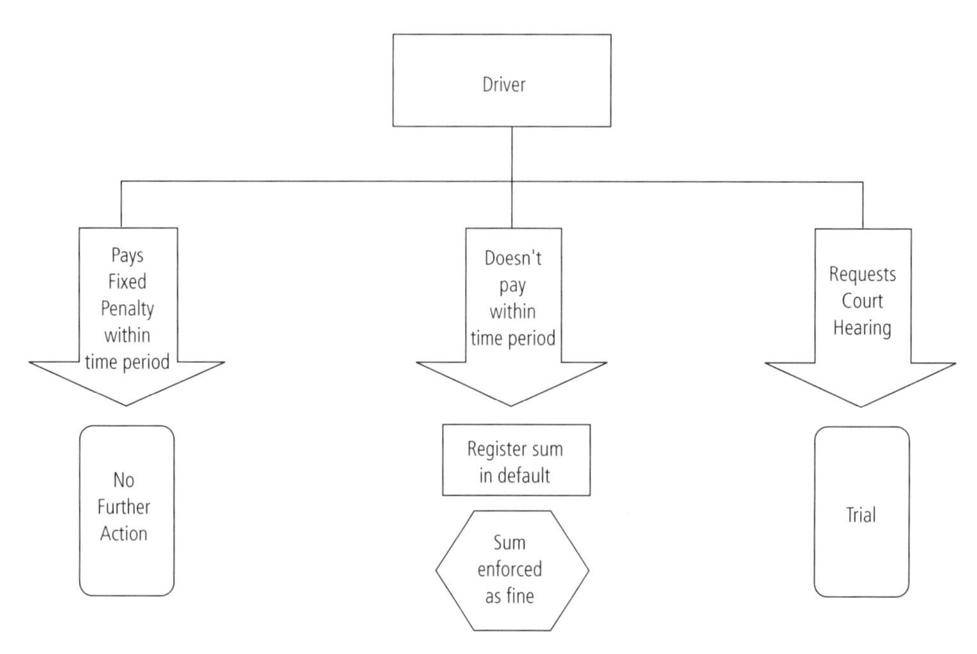

Diagram 3 Fixed Penalty Procedure: Non-Endorseable (Driver Present)

Appendix 10

Information codes

01 Eyesight correction
02 Hearing/communication aid
10 Modified transmission
15 Modified clutch
20 Modified braking systems
25 Modified accelerator systems
30 Combined braking and accelerator systems
35 Modified control layouts
40 Modified steering
42 Modified rearview mirror(s)
43 Modified driving seats
44 Modifications to motorcycles
45 Motorcycle only with sidecar
70 Exchange of licence
71 Duplicate of licence
78 Restricted to vehicles with automatic transmission
79 Restricted to vehicles in conformity with the specifications stated in brackets
101 Not for hire or reward
102 Drawbar trailers only
103 Subject to certificate of competence
105 Not more than 5.5m long
106 Restricted to vehicles with automatic transmission
107 Not more than 8250kg
108 Subject to minimum age requirements
110 Limited to invalid carriages
111 Limited to 16 passenger seats
113 Limited to 16 passenger seats except for automatics
114 With any special controls required for safe driving
115 Organ donor
118 Start date is for earliest entitlement
119 Weight limit does not apply
120 Complies with health standard for category D1

Appendix 11

Endorsement offence codes

Code	Accident Offences	Penalty Points
AC10	Failing to stop after an accident	5–10
AC20	Failing to give particulars or to report an accident within 24 hours	5–10
AC30	Undefined accident offences	4–9

Code	Disqualified Driver	Penalty Points
BA10	Driving while disqualified by order of court	6
BA30	Attempting to drive while disqualified by order of court	6

Code	Careless Driving	Penalty Points
CD10	Driving without due care and attention	3–9
CD20	Driving without reasonable consideration for other road users	3–9
CD30	Driving without due care and attention or without reasonable consideration for other road users	3–9
CD40	Causing death through careless driving when unfit through drink	3–11
CD50	Causing death by careless driving when unfit through drugs	3–11
CD60	Causing death by careless driving with alcohol level above the limit	3–11
CD70	Causing death by careless driving then failing to supply a specimen for analysis	3–11

Code	Construction and Use Offences	Penalty Points
CU10	Using a vehicle with defective brakes	3
CU20	Causing or likely to cause danger by reason of use of unsuitable vehicle or using a vehicle with parts or accessories (excluding brakes, steering or tyres) in a dangerous condition	3
CU30	Using a vehicle with defective tyre(s)	3
CU40	Using a vehicle with defective steering	3
CU50	Causing or likely to cause danger by reason of load or passengers	3

Code	Reckless/Dangerous Driving	Penalty Points
DD40	Dangerous driving	3–11
DD60	Manslaughter or culpable homicide while driving a vehicle	3–11
DD80	Causing death by dangerous driving	3–11

Code	Drink or Drugs	Penalty Points
DR10	Driving or attempting to drive with alcohol level above limit	3–11
DR20	Driving or attempting to drive while unfit through drink	3–11
DR30	Driving or attempting to drive then failing to supply a specimen for analysis	3–11
DR40	In charge of a vehicle when alcohol level above limit	10
DR50	In charge of a vehicle while unfit through drink	10
DR60	Failure to provide a specimen for analysis in circumstances other than driving or attempting to drive	10
DR70	Failing to provide specimen for breath test	4
DR80	Driving or attempting to drive when unfit through drugs	3–11
DR90	In charge of a vehicle when unfit through drugs	10

Code	Insurance Offences	Penalty Points
IN10	Using a vehicle uninsured against third party risks	6–8

Code	Licence Offences	Penalty Points
LC20	Driving otherwise than in accordance with a licence	3–6
LC30	Driving after making a false declaration about fitness when applying for a licence	3–6
LC40	Driving a vehicle having failed to notify a disability	3–6
LC50	Driving after a licence has been revoked or refused on medical grounds	3–6

Code	Miscellaneous Offences	Penalty Points
MS10	Leaving a vehicle in a dangerous position	3
MS20	Unlawful pillion riding	3
MS30	Play street offences	2
MS40	Driving with uncorrected defective eyesight or refusing to submit to a test	3
MS50	Motor racing on the highway	3–11
MS60	Offences not covered by other codes as appropriate	
MS70	Driving with uncorrected defective eyesight	3
MS80	Refusing to submit to an eyesight test	3
MS90	Failure to give information as to identity of driver etc.	3

Code	Motorway Offences	Penalty Points
MW10	Contravention of special roads Regulations (excluding speed limits)	3

Code	Pedestrian Crossings	Penalty Points
PC10	Undefined contravention of Pedestrian Crossing Regulations	3
PC20	Contravention of Pedestrian Crossing Regulations with moving vehicle	3
PC30	Contravention of Pedestrian Crossing Regulations with stationary vehicle	3

Code	Speed Limits	Penalty Points
SP10	Exceeding goods vehicle speed limits	3–6
SP20	Exceeding speed limit for type of vehicle (excluding goods or passenger vehicles)	3–6
SP30	Exceeding statutory speed limit on a public road	3–6
SP40	Exceeding passenger vehicle speed limit	3–6
SP50	Exceeding speed limit on a motorway	3–6
SP60	Undefined speed limit offence	3–6

Code	Traffic Direction and Signs	Penalty Points
TS10	Failing to comply with traffic light signals	3
TS20	Failing to comply with double white lines	3
TS30	Failing to comply with a 'stop' sign	3
TS40	Failing to comply with direction of a constable/warden	3
TS50	Failing to comply with a traffic sign (excluding 'stop' signs, traffic lights or double white lines)	3
TS60	Failing to comply with a school crossing patrol sign	3
TS70	Undefined failure to comply with a traffic direction sign	3

Code	Special Code	Penalty Points
TT99	To signify a disqualification under 'totting up' procedure. If the total of penalty points reaches 12 or more within 3 years, the driver is liable to be disqualified	

Code	Theft or Unauthorised Taking	Points
UT50	Aggravated taking of a vehicle	3–11

Appendix 12

Police (Conduct) Regulations 1999: extracts from the Home Office/ACPO Joint Guidelines

Police Officers Convicted of Drink Driving Offences

The Home Office and Police Service are committed to reducing incidents of drink driving, both generally and within the Service.

The damage done to the reputation of the Service by officers convicted of these offences cannot be overstated and detracts from the credibility of the Service in this crucial area of law enforcement.

An officer convicted by a Court of a drink driving offence can expect to face a formal disciplinary hearing. The usual sanction to be applied is either dismissal or a requirement to resign to reflect the serious view which is taken, both inside the Service and by society generally.

A Discipline panel will always treat each case on its merits but officers presiding at such hearings must apply their judgement to the facts of the case to consider whether an alternative sanction could be justified. Aggravating factors in considering the seriousness of an offence include where:

- the offence was committed on duty;
- there is an attempt to avoid arrest;
- there is an attempt to interfere with due process, particularly by leaving the scene or improperly using his position as a police officer;
- the alcohol reading is particularly high; or
- the offence derives from a traffic accident or other incident involving a member of the public.

Only in cases where none of these circumstances exist and there are exceptional circumstances should a lesser sanction be imposed. When this happens the reasons should be clearly set out and recorded.

. . .

The presiding officer must judge which is the most appropriate course of action in all the circumstances of the case. Nothing in this paragraph should be taken to suggest that, where an officer's medical condition is found to be such that he or she would normally be retired on medical grounds, the misconduct proceedings should prevent or delay retirement. However, there may be some cases, especially those where the conduct in question is very serious, where it may not be in the public or the force's interest to proceed with medical retirement in advance of a misconduct hearing, held in the absence of the officer concerned if necessary. In the event of medical retirement the misconduct proceedings will automatically lapse. . . .

V Outcomes of Hearings

3.69 There will, essentially, be three issues for the hearing to decide; what actually happened on the occasion in question, did the officer's conduct on that occasion fail to meet the standards set out in the Code of Conduct and, if so, what should be done about it? In considering the first two of these issues, the officers taking the hearing will make every effort to discover the truth and to assess the situation impartially. Their consideration of the issues before them will involve a number of factors, only some of which may require individual actions on the part of the officer concerned to be established.

3.70 In deciding matters of fact the burden of proof lies with the presenting officer, and the tribunal must apply the standard of proof required in civil cases, that is, the balance of probabilities. The straightforward legal definition of the civil standard of proof is that the adjudicator is convinced by the evidence that it is more likely or probable that something occurred than that it did not occur. Relevant case law makes it clear that the degree of proof required increases with the gravity of what is alleged and its potential consequences. It therefore follows that, where an allegation is likely to ruin an officer's reputation, deprive them of their livelihood or seriously damage their career prospects, a tribunal should be satisfied to a high degree of probability that what is alleged has been proved.

3.71 Misconduct tribunals should always bear in mind the fact that police officers are required to deal with people who may have a particular motive for making false allegations against them. An officer facing serious consequences is entitled to expect a tribunal to give very careful consideration to the evidence before an allegation is found to be proved, and tribunals should always look for other evidence which supports that given by a complainant.

3.72 Where the officers taking the hearing decide that the officer concerned has acted in a way which falls below the required standards, they will then need to consider how best to deal with the matter. In considering this question, the hearing will have regard to the officer's record of police service as shown on his or her personal record (normally with oral evidence being given by the Superintendent or Chief Inspector with management responsibility for the unit in which the officer concerned works), and will take into account any mitigating or aggravating factors.

3.73 Evidence given by the line manager will normally include details of the officer's current work performance, and care must be taken to ensure that any such details are relevant to the issue before the hearing, accurate and known to the officer. The evidence given by the line manager should cover all relevant matters, including those which are to the officer's credit, and must be fair and accurate. The officer who is to present the evidence should therefore exercise considerable care in its preparation and, in particular, should ensure that nothing is included in the evidence which cannot be substantiated. The evidence should not include details of any misconduct matters which have been expunged from the officer's personal record, details of informally resolved complaints or other misconduct matters which are not formally recorded on the officer's personal record.

3.74 In addition, the officer concerned or his or her 'friend' (or legal representative) will be given the opportunity to make representations on the question of the most appropriate way of disposing of the case and to produce witnesses as to the officer's character, whether or not witnesses have been heard in the main body of the hearing.

3.75 Where an officer has mounted a defence of public interest disclosure, the tribunal should have regard to whether the correct procedure was followed. . . .

3.76 The full range of sanctions available to a hearing is:

(a) *Dismissal*—effective immediately;

(b) *Requirement to resign*—either immediately or after such other period as the tribunal may specify;

(c) *Reduction in rank*—effective immediately;

(d) *Fine*—of not more than 13 days' pay recoverable over a minimum of 13 weeks;

(e) *Reprimand*—which would be recorded in the officer's personal record;

(f) *Caution*—which would not be recorded in the officer's personal record.

3.77 In cases involving a conviction for a drink driving offence Guidelines have been agreed with ACPO and are at Annex N [of the Guidance to Chief Officers]. They should be taken into account before a sentence is imposed in cases where there is a conviction for driving whilst over the limit.

3.78 In addition, in a case where it has been established that there was a failure to meet standards, it is open to the hearing to decide that no further action need be taken in relation to the officer concerned.

3.79 The officers taking the hearing will retire or, if necessary, adjourn the hearing (normally only overnight) in order to consider their decisions. The officer concerned will be informed of the hearing's decisions orally, in person, by the presiding officer. He or she will then be notified in writing by the presiding officer of the decisions of the hearing not more than three days after the completion of the hearing. Together with this notification, the officer will also be informed of the right to seek a Chief Constable's review of the decisions of the hearing. The notification of the decisions of the hearing will be accompanied by a copy of the presiding officer's account of the hearing's conclusions about the failure to meet standards, its views on any mitigating or aggravating factors, and the reasons for the decisions of the hearing as to finding and outcome.

3.80 If a hearing is adjourned the officers conducting it should not discuss the details of the case with anyone else, whether connected with the case or not. It is only in this way that the impartiality of the proceedings can be seen to be protected adequately.

VI Avenues of Appeal

Chief Constable's review

3.81 An officer has the right to require the decisions of the hearing to be reviewed by his or her Chief Constable. The request for a review should be made within 14 days of the written notification of the hearing's decision, and should be accompanied by written grounds in support of the request....

Appeal to Police Appeals Tribunal

3.82 An officer who has been dismissed, required to resign or reduced in rank following a hearing and the Chief Constable's review will have a right of appeal to a Police Appeals Tribunal....

VII Records

3.83 Where an officer is found not to have fallen below the standards set out in the Code of Conduct, or receives a caution from a hearing, or the hearing decides to take no action, no entry should be made in the officer's service record. Where a record is made of the outcome of any misconduct hearing it will be expunged from the officer's service record after a period of three or five years, depending on the punishment imposed (regulation 17(2) of the Police Regulations 1995 refers). Records of written warnings will be expunged after 12 months. Where written warnings or the outcome of any misconduct hearings have been expunged from an officer's service record they should not be referred to in any future misconduct proceedings.

Misconduct books

3.84 Regulations require that each force keeps a record of misconduct hearings held to deal with alleged breaches of the Code of Conduct, including the outcome of any such action. Forces may also wish to keep records of other action taken which falls short of formal proceedings.

Appendix 13

Vehicles Crime (Registration of Registration Plate Suppliers) (England and Wales) Regulations 2002, Schedule

Regulation 6

SCHEDULE 2
PART 1

Documents which may be used to verify name and address

1. The document used shall be either that described in paragraph 2, or one of the documents described in paragraph 3 together with one of the documents described in paragraph 4.

2. The document described in this paragraph is a driving licence which includes a photograph of the holder (whether or not issued in the United Kingdom).

3. The documents described in this paragraph are in each case addressed to the purchaser or individual acting for him and dated within 6 months of the purchase—

(a) a bill issued by an electricity, gas or water supplier,

(b) a bank or building society statement.

4. The documents described in this paragraph are—

(a) a driving licence whether or not issued in the United Kingdom (without a photograph of the holder),

(b) a passport (whether or not issued by the United Kingdom),

(c) a national identity card issued by the government of a state or territory other than the United Kingdom,

(d) a debit or credit card having a photograph of the holder issued by a bank or building society,

(e) a travel pass having a photograph of the holder issued by a local authority, a person having an operator's licence issued under Part II of the Public Passenger Vehicles Act 1981 or a passenger service operator as defined in section 83(1) of the Railways Act 1993.

PART II

Documents which may be used to establish connection with the vehicle

The document used shall be one of the following—

(1) V5 (registration document),

(2) V750 (certificate of entitlement to a registration mark),

(3) V778 (right to a particular registration mark under section 27 of the Vehicles Excise and Registration Act 1994),

(4) V11 (vehicle licensing reminder issued to registered keeper by the Secretary of State),

(5) V379 (certificate of vehicle registration),

(6) V948 (authorisation issued by the Secretary of State for the purchase of the number plate),

(7) An authorisation for the purchase of the number plate issued by a company owning more than one vehicle stating that it holds the V5 and giving either the reference number from the V5 or the vehicle identification number.

Index